THE FOUNDATIONS OF ARITHMETIC

The

Foundations of Arithmetic

A logico-mathematical enquiry into the concept of number

GOTTLOB FREGE

ENGLISH TRANSLATION BY
J. L. AUSTIN, M.A.
*Late White's Professor of Moral
Philosophy in the University of Oxford*

Second Revised Edition

BASIL BLACKWELL
OXFORD
1980

© *Basil Blackwell Publisher, 1950, 1953, 1980*

First printed in 1950
Second edition, 1953
Reprinted 1959, 1968, 1974, 1978
Fifth impression with corrections, 1980
English language only edition
first published in paperback, 1980

ISBN 0 631 12694 5

Translator's Note

Words and footnotes in square brackets are insertions by the translator.

The pagination is the same as in the original German edition, except that in the original the "Analysis of Contents" pages were not numbered.

Some of Frege's references and quotations, which are not always accurate, have been corrected in the translated version.

Translator's Preface to the Second Edition

Though users of the first edition of this version will perhaps not be anywhere seriously misled in doctrine, a large number of passages in it have called, and some even howled, for improvements in fidelity or lucidity. The translator's thanks are due to several readers, and in particular to Mr. P. T. Geach, for their trouble in contributing emendations and suggestions: nothing could be more welcome than more of the same.*

There is justice in the general criticism that the version is too long. Here and there it has been possible to do something to correct this, but it is too late and too difficult now to strike a fresh compromise throughout between the claims of brevity and those of naturalness and clarity. Frege is an unusually, even at times an unduly, succinct writer, and the German text must be allowed to remain the final testimony to his style.

The translations originally chosen for Frege's principal terms remain unchanged, except that *Begriffswort* has now become "concept word" instead of "general term" and *wirklich* "actual" instead of "existent." Critics of some others of these translations have perhaps not sufficiently realized that Frege's inherited philosophical vocabulary (at least as he was using it at this period) is a dated one. It is that which was Englished by his contemporaries, the "British Idealists": and they certainly used, for example, "idea" for *Vorstellung* and "proposition" for *Satz*, though not unnaturally they attached to those words meanings different from (and doubtless less clear than) those fashionable half a century later. Frege's thought cannot be reproduced accurately, nor can his terms be translated consistently, unless we are prepared to accept, even in him, something short of complete (or contemporary) sophistication.

*) *Note:* The distinguished translator died in 1960. In the 1980 impression a few amendments such as he called for have been made at the suggestion of Professor M. A. E. Dummett and Mr B. F. McGuinness.

Analysis of Contents.

I. Views of certain writers on the nature of arithmetical propositions.

Are numerical formulae provable?

II. Views of certain writers on the concept of Number.

III. Views on unity and one.

Does the number word "one" express a property of objects?

Are units identical with one another?

IV. The concept of Number.

Every individual number is a self-subsistent object.

To obtain the concept of Number, we must fix the sense of a numerical identity.

Our definition completed and its worth proved.

PAGE

INTRODUCTION

When we ask someone what the number one is, or what the symbol 1 means,* we get as a rule the answer "Why, a thing". And if we go on to point out that the proposition

"the number one is a thing"

is not a definition, because it has the definite article on one side and the indefinite on the other, or that it only assigns the number one to the class of things, without stating which thing it is, then we shall very likely be invited to select something for ourselves—anything we please—to call one. Yet if everyone had the right to understand by this name whatever he pleased, then the same proposition about one would mean different things for different people,—such propositions would have no common content. Some, perhaps, will decline to answer the question, pointing out that it is impossible to state, either, what is meant by the letter a, as it is used in arithmetic; and that if we were to say "a means a number," this would be open to the same objection as the definition "one is a thing." Now in the case of a it is quite right to decline to answer: a does not mean some one definite number which can be specified, but serves to express the generality of general propositions. If, in $a + a - a = a$, we put for a

* [I have tried throughout to translate *Bedeutung* and its cognates by "meaning" and *Sinn* and its cognates by "sense", in view of the importance Frege later attached to the distinction. But it is quite evident that he attached no special significance to the words at this period.]

some number, any we please but the same throughout, we get always a true identity.* This is the sense in which the letter *a* is used. With one, however, the position is essentially different. Can we, in the identity $1 + 1 = 2$, put for 1 in both places some one and the same object, say the Moon? On the contrary, it looks as though, whatever we put for the first 1, we must put something different for the second. Why is it that we have to do here precisely what would have been wrong in the other case? Again, arithmetic cannot get along with *a* alone, but has to use further letters besides (*b*, *c* and so on), in order to express in general form relations between different numbers. It would therefore be natural to suppose that the symbol 1 too, if it served in some similar way to confer generality on propositions, could not be enough by itself. Yet surely the number one looks like a definite particular object, with properties that can be specified, for example that of remaining unchanged when multiplied by itself? In this sense, *a* has no properties that can be specified, since whatever can be asserted of *a* is a common property of all numbers, whereas $1^1 = 1$ asserts nothing of the Moon, nothing of the Sun, nothing of the Sahara, nothing of the Peak of Teneriffe; for what could be the sense of any such assertion?

Questions like these catch even mathematicians for that matter, or most of them, unprepared with any satisfactory answer. Yet is it not a scandal that our science should be so unclear about the first and foremost among its objects, and one which is apparently so simple? Small hope, then, that we shall be able to say what number is. If a concept fundamental to a mighty science gives rise to difficulties, then it is surely an imperative task to investigate it more closely until those difficulties are overcome; especially as we shall hardly succeed in finally clearing up negative numbers, or fractional or complex numbers, so long as our insight into the foundation of the whole structure of arithmetic is still defective.

* [*Gleichung*. This also means, and would often be more naturally translated, "equation". But I have generally retained "identity", because this is sometimes essential and because Frege does understand equations as identities. For similar reasons I have translated *gleich* "identical", though it can mean "equal" or even merely "similar". Cp. §§ 34, 65.]

Admittedly, many people will think this not worth the trouble. Naturally, they suppose, this concept is adequately dealt with in the elementary textbooks, where the subject is settled once and for all. Who can believe that he has anything still to learn on so simple a matter? So free from all difficulty is the concept of positive whole number held to be, that an account of it fit for children can be both scientific and exhaustive; and that every schoolboy, without any further reflexion or acquaintance with what others have thought, knows all there is to know about it. The first prerequisite for learning anything is thus utterly lacking—I mean, the knowledge that we do not know. The result is that we still rest content with the crudest of views, even though since HERBART'S[1] day a better doctrine has been available. It is sad and discouraging to observe how discoveries once made are always threatening to be lost again in this way, and how much work promises to have been done in vain, because we fancy ourselves so well off that we need not bother to assimilate its results. My work too, as I am well aware, is exposed to this risk. A typical crudity confronts me, when I find calculation described as "aggregative mechanical thought".[2] I doubt whether there exists any thought whatsoever answering to this description. An aggregative imagination, even, might sooner be let pass; but that has no relevance to calculation. Thought is in essentials the same everywhere: it is not true that there are different kinds of laws of thought to suit the different kinds of objects thought about. Such differences as there are consist only in this, that the thought is more pure or less pure, less dependent or more upon psychological influences and on external aids such as words or numerals, and further to some

---

[1] Collected Works, ed. Hartenstein, Vol. X, part i, *Umriss pädagogischer Vorlesungen*, § 252, n. 2: "Two does not mean two things, but doubling" etc.

[2] K. Fischer, *System der Logik und Metaphysik oder Wissenschaftslehre*, 2nd edn., §94.

extent too in the finer or coarser structure of the concepts involved; but it is precisely in this respect that mathematics aspires to surpass all other sciences, even philosophy.

The present work will make it clear that even an inference like that from n to $n + 1$, which on the face of it is peculiar to mathematics, is based on the general laws of logic, and that there is no need of special laws for aggregative thought. It is possible, of course, to operate with figures mechanically, just as it is possible to speak like a parrot: but that hardly deserves the name of thought. It only becomes possible at all after the mathematical notation has, as a result of genuine thought, been so developed that it does the thinking for us, so to speak. This does not prove that numbers are formed in some peculiarly mechanical way, as sand, say, is formed out of quartz granules. In their own interests mathematicians should, I consider, combat any view of this kind, since it is calculated to lead to the disparagement of a principal object of their study, and of their science itself along with it. Yet even in the works of mathematicians are to be found expressions of exactly the same sort. The truth is quite the other way: the concept of number, as we shall be forced to recognize, has a finer structure than most of the concepts of the other sciences, even although it is still one of the simplest in arithmetic.

In order, then, to dispel this illusion that the positive whole numbers really present no difficulties at all, but that universal concord reigns about them, I have adopted the plan of criticizing some of the views put forward by mathematicians and philosophers on the questions involved. It will be seen how small is the extent of their agreement—so small, that we find one dictum precisely contradicting another. For example, some hold that "units are identical with one another," others that they are different, and each side supports its assertion with arguments that cannot be rejected out of hand. My object in this is to awaken a desire for a stricter enquiry. At the same time this preliminary examination of the views others have put forward should clear the ground for my own account, by convincing my readers in advance that these other paths do not lead to the goal, and that my opinion is not just one among many all equally tenable; and in this way I hope to settle the question finally, at least in essentials.

I realize that, as a result, I have been led to pursue arguments more philosophical than many mathematicians may approve; but any thorough investigation of the concept of number is bound always to turn out rather philosophical. It is a task which is common to mathematics and philosophy.

It may well be that the co-operation between these two sciences, in spite of many démarches from both sides, is not so flourishing as could be wished and would, for that matter, be possible. And if so, this is due in my opinion to the predominance in philosophy of psychological methods of argument, which have penetrated even into the field of logic. With this tendency mathematics is completely out of sympathy, and this easily accounts for the aversion to philosophical arguments felt by many mathematicians. When STRICKER,[1] for instance, calls our ideas* of numbers motor phenomena and makes them dependent on muscular sensations, no mathematician can recognize his numbers in such stuff or knows what on earth to make such a proposition. An arithmetic founded on muscular sensations would certainly turn out sensational enough, but also every bit as vague as its foundation. No, sensations are absolutely no concern of arithmetic. No more are mental pictures, formed from the amalgamated traces of earlier sense-impressions. All these phases of consciousness are characteristically fluctuating and indefinite, in strong contrast to the definiteness and fixity of the concepts and objects of

---

[1] *Studien über Association der Vorstellungen*, Vienna 1883.

* [*Vorstellungen*. I have translated this word consistently by "idea", and cognate words by "imagine", "imagination", etc. For Frege it is a psychological term, cp. p. x^e below.]

mathematics. It may, of course, serve some purpose to investigate the ideas and changes of ideas which occur during the course of mathematical thinking; but psychology should not imagine that it can contribute anything whatever to the foundation of arithmetic. To the mathematician as such these mental pictures, with their origins and their transformations, are immaterial. STRICKER himself states that the only idea he associates with the word "hundred" is the symbol 100. Others may have the idea of the letter C or something else; does it not follow, therefore, that these mental pictures are, so far as concerns us and the essentials of our problem, completely immaterial and incidental—as incidental as chalk and blackboard, and indeed that they do not deserve to be called ideas of the number a hundred at all? Never, then, let us suppose that the essence of the matter lies in such ideas. Never let us take a description of the origin of an idea for a definition, or an account of the mental and physical conditions on which we become conscious of a proposition for a proof of it. A proposition may be thought, and again it may be true; let us never confuse these two things. We must remind ourselves, it seems, that a proposition no more ceases to be true when I cease to think of it than the sun ceases to exist when I shut my eyes. Otherwise, in proving Pythagoras' theorem we should be reduced to allowing for the phosphorous content of the human brain; and astronomers would hesitate to draw any conclusions about the distant past, for fear of being charged with anachronism,—with reckoning twice two as four regardless of the fact that our idea of number is a product of evolution and has a history behind it. It might be doubted whether by that time it had progressed so far. How could they profess to know that the proposition $2 \times 2 = 4$ already held good in that remote epoch? Might not the creatures then extant have held the proposition $2 \times 2 = 5$, from which the proposition $2 \times 2 = 4$ was only evolved later through a process of natural selection in the struggle for existence? Why, it might even be that $2 \times 2 = 4$ itself is destined in the same way to develop into $2 \times 2 = 3$! *Est modus in rebus, sunt certi denique fines!* The historical approach, with its aim of detecting how things begin and of arriving from these origins at a knowledge of their nature, is certainly perfectly legitimate; but it has also its limitations. If everything were in continual flux, and nothing maintained itself fixed for all time, there would no longer be any possibility of getting to know anything about the world and everything would be plunged in confusion. We suppose, it would seem, that concepts sprout in the individual mind like leaves on a tree, and we think to discover their nature by studying their birth: we seek to define them psychologically, in terms of the nature of the human mind. But this account makes everything subjective, and if we follow it through to the end, does away with truth. What is known as the history of concepts is really a history either of our knowledge of concepts or of the meanings of words. Often it is only after immense intellectual effort, which may have continued over centuries, that humanity at last succeeds in achieving knowledge of a concept in its pure form, in stripping off the irrelevant accretions which veil it from the eyes of the mind. What, then, are we to say of those who, instead of advancing this work where it is not yet completed, despise it, and betake themselves to the nursery, or bury themselves in the remotest conceivable periods of human evolution, there to discover, like JOHN STUART MILL, some gingerbread or pebble arithmetic! It remains only to ascribe to the flavour of the bread some special meaning for the concept of number. A procedure like this is surely the very reverse of rational, and as unmathematical, at any rate, as it could well be. No wonder the mathematicians turn their backs on it. Do the concepts, as we approach their supposed sources, reveal themselves in peculiar purity? Not at all;

we see everything as through a fog, blurred and undifferentiated. It is as though everyone who wished to know about America were to try to put himself back in the position of Columbus, at the time when he caught the first dubious glimpse of his supposed India. Of course, a comparison like this proves nothing; but it should, I hope, make my point clear. It may well be that in many cases the history of earlier discoveries is a useful study, as a preparation for further researches; but it should not set up to usurp their place.

So far as mathematicians are concerned, an attack on such views would indeed scarcely have been necessary; but my treatment was designed to bring each dispute to an issue for the philosophers as well, as far as possible, so that I found myself forced to enter a little into psychology, if only to repel its invasion of mathematics.

Besides, even mathematical textbooks make use of psychological expressions. When the author feels himself obliged to give a definition, yet cannot, then he tends to give at least a description of the way in which we arrive at the object or concept concerned. These cases can easily be recognized by the fact that such explanations are never referred to again in the course of the subsequent exposition. For teaching purposes, introductory devices are certainly quite legitimate; only they should always be clearly distinguished from definitions. A delightful example of the way in which even mathematicians can confuse the grounds of proof with the mental or physical conditions to be satisfied if the proof is to be given is to be found in E. SCHRÖDER.[1] Under the heading "Special Axiom" he produces the following: "The principle I have in mind might well be called the Axiom of Symbolic Stability. It guarantees us that throughout all our arguments and deductions the symbols

[1] *Lehrbuch der Arithmetik und Algebra*, [Leipzig 1873].

remain constant in our memory—or preferably on paper," and so on.

No less essential for mathematics than the refusal of all assistance from the direction of psychology, is the recognition of its close connexion with logic. I go so far as to agree with those who hold that it is impossible to effect any sharp separation of the two. This much everyone would allow, that any enquiry into the cogency of a proof or the justification of a definition must be a matter of logic. But such enquiries simply cannot be eliminated from mathematics, for it is only through answering them that we can attain to the necessary certainty.

In this direction too I go, certainly, further than is usual. Most mathematicians rest content, in enquiries of this kind, when they have satisfied their immediate needs. If a definition shows itself tractable when used in proofs, if no contradictions are anywhere encountered, and if connexions are revealed between matters apparently remote from one another, this leading to an advance in order and regularity, it is usual to regard the definition as sufficiently established, and few questions are asked as to its logical justification. This procedure has at least the advantage that it makes it difficult to miss the mark altogether. Even I agree that definitions must show their worth by their fruitfulness: it must be possible to use them for constructing proofs. Yet it must still be borne in mind that the rigour of the proof remains an illusion, even though no link be missing in the chain of our deductions, so long as the definitions are justified only as an afterthought, by our failing to come across any contradiction. By these methods we shall, at bottom, never have achieved more than an empirical certainty, and we must really face the possibility that we may still in the end encounter a contradiction which brings the whole edifice down in ruins. For this reason I have felt bound to go back rather further into the general logical foundations of our science than perhaps most mathematicians will consider necessary.

In the enquiry that follows, I have kept to three fundamental principles:

always to separate sharply the psychological from the
logical, the subjective from the objective;

never to ask for the meaning of a word in isolation, but
only in the context of a proposition;

never to lose sight of the distinction between concept
and object.

In compliance with the first principle, I have used the word
"idea" always in the psychological sense, and have distinguished
ideas from concepts and from objects. If the second principle
is not observed, one is almost forced to take as the meanings
of words mental pictures or acts of the individual mind, and
so to offend against the first principle as well. As to the third
point, it is a mere illusion to suppose that a concept can be
made an object without altering it. From this it follows that a
widely-held formalist theory of fractional, negative, etc., numbers is untenable. How I propose to improve upon it can be no
more than indicated in the present work. With numbers of
all these types, as with the positive whole numbers, it is a
matter of fixing the sense of an identity.

My results will, I think, at least in essentials, win the
adherence of those mathematicians who take the trouble to
attend to my arguments. They seem to me to be in the air,
and it may be that every one of them singly, or at least something very like it, has been already put forward; though
perhaps, presented as they are here in connexion with each
other, they may still be novel. I have often been astonished
at the way in which writers who on one point approach my
view so closely, on others depart from it so violently.

Their reception by philosophers will be varied, depending
on each philosopher's own position; but presumably those

empiricists who recognize induction as the sole original process of inference (and even that as a process not actually of inference but of habituation) will like them least. Some one or another, perhaps, will take this opportunity to examine afresh the principles of his theory of knowledge. To those who feel inclined to criticize my definitions as unnatural, I would suggest that the point here is not whether they are natural, but whether they go to the root of the matter and are logically beyond criticism.

I permit myself the hope that even the philosophers, if they examine what I have written without prejudice, will find in it something of use to them.

§ 1. After deserting for a time the old Euclidean standards of rigour, mathematics is now returning to them, and even making efforts to go beyond them. In arithmetic, if only because many of its methods and concepts originated in India, it has been the tradition to reason less strictly than in geometry, which was in the main developed by the Greeks. The discovery of higher analysis only served to confirm this tendency; for considerable, almost insuperable, difficulties stood in the way of any rigorous treatment of these subjects, while at the same time small reward seemed likely for the efforts expended in overcoming them. Later developments, however, have shown more and more clearly that in mathematics a mere moral conviction, supported by a mass of successful applications, is not good enough. Proof is now demanded of many things that formerly passed as self-evident. Again and again the limits to the validity of a proposition have been in this way established for the first time. The concepts of function, of continuity, of limit and of infinity have been shown to stand in need of sharper definition. Negative and irrational numbers, which had long since been admitted into science, have had to submit to a closer scrutiny of their credentials.

In all directions these same ideals can be seen at work—rigour of proof, precise delimitation of extent of validity, and as a means to this, sharp definition of concepts.

§ 2. Proceeding along these lines, we are bound even-
tually to come to the concept of Number* and to the simplest
propositions holding of positive whole numbers, which form
the foundation of the whole of arithmetic. Of course, numerical
formulae like $7 + 5 = 12$ and laws like the Associative Law
of Addition are so amply established by the countless applications made of them every day, that it may seem almost ridicu-
lous to try to bring them into dispute by demanding a proof
of them. But it is in the nature of mathematics always to
prefer proof, where proof is possible, to any confirmation by
induction. Euclid gives proofs of many things which anyone would concede him without question. And it was when
men refused to be satisfied even with Euclid's standards of
rigour that they were led to the enquiries set in train by the
Axiom of Parallels.

Thus our movement in favour of all possible rigour has
already outstripped in many directions the demand originally
raised, while the demand has itself continually grown in scope
and urgency.

The aim of proof is, in fact, not merely to place the truth
of a proposition beyond all doubt, but also to afford us insight
into the dependence of truths upon one another. After we have
convinced ourselves that a boulder is immovable, by trying
unsuccessfully to move it, there remains the further question,
what is it that supports it so securely? The further we pursue
these enquiries, the fewer become the primitive truths to
which we reduce everything; and this simplification is in itself
a goal worth pursuing. But there may even be justification for
a further hope: if, by examining the simplest cases, we can
bring to light what mankind has there done by instinct,
and can extract from such procedures what is universally valid
in them, may we not thus arrive at general methods for forming
concepts and establishing principles which will be applicable
also in more complicated cases?

* [*Anzahl*, i.e. cardinal number, cp. § 4 n. I have always used "Number"
to translate this, and "number" for the more usual and general *Zahl*. Throughout most of the present work the distinction is not important, and Frege uses
the two words almost indifferently.]

§ 3. Philosophical motives too have prompted me to enquiries of this kind. The answers to the questions raised about the nature of arithmetical truths—are they a priori or a posteriori? synthetic or analytic?—must lie in this same direction. For even though the concepts concerned may themselves belong to philosophy, yet, as I believe, no decision on these questions can be reached without assistance from mathematics—though this depends of course on the sense in which we understand them.

It not uncommonly happens that we first discover the content of a proposition, and only later give the rigorous proof of it, on other and more difficult lines; and often this same proof also reveals more precisely the conditions restricting the validity of the original proposition. In general, therefore, the question of how we arrive at the content of a judgement should be kept distinct from the other question, Whence do we derive the justification for its assertion?

Now these distinctions between a priori and a posteriori, synthetic and analytic, concern, as I see it,[1] not the content of the judgement but the justification for making the judgement. Where there is no such justification, the possibility of drawing the distinctions vanishes. An a priori error is thus as complete a nonsense as, say, a blue concept. When a proposition is called a posteriori or analytic in my sense, this is not a judgement about the conditions, psychological, physiological and physical, which have made it possible to form the content of the proposition in our consciousness; nor is it a judgement about the way in which some other man has come, perhaps erroneously, to believe it true; rather, it is a judgement about the ultimate ground upon which rests the justification for holding it to be true.

This means that the question is removed from the sphere of psychology, and assigned, if the truth concerned is a

---

[1] By this I do not, of course, mean to assign a new sense to these terms, but only to state accurately what earlier writers, KANT in particular, have meant by them.

mathematical one, to the sphere of mathematics. The problem becomes, in fact, that of finding the proof of the proposition, and of following it up right back to the primitive truths. If, in carrying out this process, we come only on general logical laws and on definitions, then the truth is an analytic one, bearing in mind that we must take account also of all propositions upon which the admissibility of any of the definitions depends. If, however, it is impossible to give the proof without making use of truths which are not of a general logical nature, but belong to the sphere of some special science, then the proposition is a synthetic one. For a truth to be a posteriori, it must be impossible to construct a proof of it without including an appeal to facts, i.e., to truths which cannot be proved and are not general, since they contain assertions about particular objects. But if, on the contrary, its proof can be derived exclusively from general laws, which themselves neither need nor admit of proof, then the truth is a priori.[1]

§ 4. Starting from these philosophical questions, we are led to formulate the same demand as that which had arisen independently in the sphere of mathematics, namely that the fundamental propositions of arithmetic should be proved, if in any way possible, with the utmost rigour; for only if every gap in the chain of deductions is eliminated with the greatest care can we say with certainty upon what primitive truths the proof depends; and only when these are known shall we be able to answer our original questions.

---

[1] If we recognize the existence of general truths at all, we must also admit the existence of such primitive laws, since from mere individual facts nothing follows, unless it be on the strength of a law. Induction itself depends on the general proposition that the inductive method can establish the truth of a law, or at least some probability for it. If we deny this, induction becomes nothing more than a psychological phenomenon, a procedure which induces men to believe in the truth of a proposition, without affording the slightest justification for so believing.

If we now try to meet this demand, we very soon come to propositions which cannot be proved so long as we do not succeed in analysing concepts which occur in them into simpler concepts or in reducing them to something of greater generality. Now here it is above all Number which has to be either defined or recognized as indefinable. This is the point which the present work is meant to settle.[1] On the outcome of this task will depend the decision as to the nature of the laws of arithmetic.

To my attack on these questions themselves I shall preface something which may give a pointer towards their answers. For suppose there should prove to be grounds from other points of view for believing that the fundamental principles of arithmetic are analytic, then these would tell also in favour of their being provable and the concept of Number definable; while any grounds for believing the same truths to be a posteriori would tell in the opposite direction. The rival theories here, therefore, may well be submitted first to a passing scrutiny.

I. Views of certain writers on the nature of arithmetical propositions.

Are numerical formulae provable?

§ 5. We must distinguish numerical formulae, such as $2 + 3 = 5$, which deal with particular numbers, from general laws, which hold good for all whole numbers.

The former are held by some philosophers[2] to be unprovable and immediately self-evident like axioms. KANT[3]

---

[1] In what follows, therefore, unless special notice is given, the only "numbers" under discussion are the positive whole numbers, which give the answer to the question "How many?".

[2] Hobbes, Locke, Newton. Cf. Baumann, *Die Lehren von Zeit, Raum und Mathematik*, [Berlin 1868, Vol. I], pp. 241–42, 365 ff., 475–76. [Hobbes, *Examinatio et Emendatio Mathematicae Hodiernae*, Amsterdam 1668, Diall. I–III, esp. I, p. 19 and III, pp. 62–63; Locke, *Essay*, Bk. IV, esp. Cap. iv, § 6 and cap. vii, §§ 6 and 10; Newton, *Arithmetica Universalis*, Vol. I, cap. i–iii, esp. iii, n. 24.]

[3] *Critique of Pure Reason*; Collected Works, ed. Hartenstein, Vol. III, p. 157 [Original edns. A 164/B205].

declares them to be unprovable and synthetic, but hesitates to call them axioms because they are not general and because the number of them is infinite. HANKEL[1] justifiably calls this conception of infinitely numerous unprovable primitive truths incongruous and paradoxical. The fact is that it conflicts with one of the requirements of reason, which must be able to embrace all first principles in a survey. Besides, is it really self-evident that

$$135664 + 37863 = 173527?$$

It is not; and KANT actually urges this as an argument for holding these propositions to be synthetic. Yet it tells rather against their being unprovable; for how, if not by means of a proof, are they to be seen to be true, seeing that they are not immediately self-evident? KANT thinks he can call on our intuition of fingers or points for support, thus running the risk of making these propositions appear to be empirical, contrary to his own expressed opinion; for whatever our intuition of 37863 fingers may be, it is at least certainly not pure. Moreover, the term "intuition" seems hardly appropriate, since even 10 fingers can, in different arrangements, give rise to very different intuitions. And have we, in fact, an intuition of 135664 fingers or points at all? If we had, and if we had another of 37863 fingers and a third of 173527 fingers, then the correctness of our formula, if it were unprovable, would have to be evident right away, at least as applying to fingers; but it is not.

KANT, obviously, was thinking only of small numbers. So that for large numbers the formulae would be provable, though for small numbers they are immediately self-evident through intuition. Yet it is awkward to make a fundamental distinction between small and large numbers, especially as it would scarcely be possible to draw any sharp boundary between them. If the numerical formulae were provable

---

[1] *Vorlesungen über die complexen Zahlen und ihren Functionen*, p. 53.

from, say, 10 on, we should ask with justice "Why not from 5 on? or from 2 on? or from 1 on?"

§ 6. Other philosophers again, and mathematicians, have asserted that numerical formulae are actually provable. LEIBNIZ[1] says:

"It is not an immediate truth that 2 and 2 are 4; provided it be granted that 4 signifies 3 and 1. It can be proved, as follows:

Definitions: (1) 2 is 1 and 1
(2) 3 is 2 and 1
(3) 4 is 3 and 1

Axiom: If equals be substituted for equals, the equality remains.*

Proof: $2 + 2 = 2 + 1 + 1$ (by Def. 1) $= 3 + 1$ (by Def. 2) $= 4$ (by Def. 3).

∴ $2 + 2 = 4$ (by the Axiom)."

This proof seems at first sight to be constructed entirely from definitions and the axiom cited. And the axiom too could be transformed into a definition, as LEIBNIZ himself does transform it in another passage.[2] It seems as though we need to know no more of 1, 2, 3 and 4 than is contained in the definitions. If we look more closely, however, we can discover a gap in the proof, which is concealed owing to the omission of the brackets. To be strictly accurate, that is, we should have to write:

$$2 + 2 = 2 + (1 + 1)$$
$$(2 + 1) + 1 = 3 + 1 = 4.$$

What is missing here is the proposition

$$2 + (1 + 1) = (2 + 1) + 1,$$

which is a special case of

$$a + (b + c) = (a + b) + c.$$

If we assume this law, it is easy to see that a similar proof can

---

[1] *Nouveaux Essais*, IV, § 10 (Erdmann edn., p. 363).

[2] *Non inelegans specimen demonstrandi in abstractis* (Erdmann edn., p. 94).

* [*Mettant des choses égales à la place, l'égalité demeure.*]

be given for every formula of addition. Every number, that means, is to be defined in terms of its predecessor. And actually I do not see how a number like 437986 could be given to us more aptly than in the way LEIBNIZ does it. Even without having any idea of it, we get it by this means at our disposal none the less. Through such definitions we reduce the whole infinite set of numbers to the number one and increase by one, and every one of the infinitely many numerical formulae can be proved from a few general propositions.

This opinion is shared by H. GRASSMANN and H. HANKEL. GRASSMANN attempts to obtain the law

$$a + (b + 1) = (a + b) + 1$$

by means of a definition, as follows[1]:

"If a and b are any arbitrary members of the basic series, then by the sum $a + b$ is to be understood that member of the basic series for which the formula

$$a + (b + e) = a + b + e$$

is valid."

e here is to be taken to mean positive unity. This definition can be criticized in two different ways. First, sum is defined in terms of itself. If we do not yet understand the meaning of $a + b$, we do not understand the expression $a + (b + e)$ either. This criticism, however, can perhaps be evaded if we say (admittedly going against the text) that what he is intending to define is not sum but addition. In that case, the criticism could still be brought that $a + b$ would be an empty symbol if there were either no member or several members of the basic series which satisfied the prescribed condition. That this does not in fact ever happen, GRASSMANN simply assumes without proof, so that the rigour of his procedure is only apparent.

[1] *Lehrbuch der Mathematik für höhere Lehranstalten*, Part i *Arithmetik*, p. 4. Stettin 1860 [= *ges. Math. u. Phys. Werke*, ed. Engel, II, i, p. 301].

§ 7. It might well be supposed that numerical formulae would be synthetic or analytic, a posteriori or a priori, according as the general laws on which their proofs depend are so. JOHN STUART MILL, however, is of the opposite opinion. At first, indeed, he seems to mean to base the science, like LEIBNIZ, on definitions,[1] since he defines the individual numbers in the same way as LEIBNIZ; but this spark of sound sense is no sooner lit than extinguished, thanks to his preconception that all knowledge is empirical. He informs us, in fact,[2] that these definitions are not definitions in the logical sense; not only do they fix the meaning of a term, but they also assert along with it an observed matter of fact. But what in the world can be the observed fact, or the physical fact (to use another of MILL's expressions), which is asserted in the definition of the number 777864? Of all the whole wealth of physical facts in his apocalypse, MILL names for us only a solitary one, the one which he holds is asserted in the definition of the number 3. It consists, according to him, in this, that collections of objects exist, which while they impress the senses thus, $^{\circ}_{\circ}{}^{\circ}$, may be separated into two parts, thus, ○○ ○. What a mercy, then, that not everything in the world is nailed down; for if it were, we should not be able to bring off this separation, and 2 + 1 would not be 3! What a pity that MILL did not also illustrate the physical facts underlying the numbers 0 and 1!

"This proposition being granted," MILL goes on, "we term all such parcels Threes." From this we can see that it is really incorrect to speak of three strokes when the clock strikes three, or to call sweet, sour and bitter three sensations

---

[1] *System of Logic*, Bk. III, cap. xxiv, § 5 (German translation by J. Schiel).
[2] Op. cit., Bk. II, cap. vi, § 2.

of taste; and equally unwarrantable is the expression "three methods of solving an equation." For none of these is a parcel which ever impresses the senses thus, $^\circ{}_\circ{}^\circ$.

Now according to Mill "the calculations do not follow from the definition itself but from the observed matter of fact." But at what point then, in the proof given above of the proposition $2 + 2 = 4$, ought Leibniz to have appealed to the fact in question? Mill omits to point out the gap in the proof, although he gives himself a precisely analogous proof of the proposition $5 + 2 = 7$.[1] Actually, there is a gap, consisting in the omission of the brackets; but Mill overlooks this just as Leibniz does.

If the definition of each individual number did really assert a special physical fact, then we should never be able sufficiently to admire, for his knowledge of nature, a man who calculates with nine-figure numbers. Meantime, perhaps Mill does not mean to go so far as to maintain that all these facts would have to be observed severally, but thinks it would be enough if we had derived through induction a general law in which they were all included together. But try to formulate this law, and it will be found impossible. It is not enough to say: "There exist large collections of things which can be split up." For this does not state that there exist collections of such a size and of such a sort as are required for, say, the number 1,000,000, nor is the manner in which they are to be divided up specified any more precisely. Mill's theory must necessarily lead to the demand that a fact should be observed specially for each number, for in a general law precisely what is peculiar to the number 1,000,000, which necessarily belongs to its definition, would be lost. On Mill's view we could actually not put 1,000,000 = 999,999 + 1 unless

[1] Op. cit., Bk. III, cap. xxiv, § 5.

we had observed a collection of things split up in precisely this peculiar way, different from that characteristic of any and every other number whatsoever.

§ 8. MILL seems to hold that we ought not to form the definitions $2 = 1 + 1$, $3 = 2 + 1$, $4 = 3 + 1$, and so on, unless and until the facts he refers to have been observed. It is quite true that we ought not to define 3 as $(2 + 1)$, if we attach no sense at all to $(2 + 1)$. But the question is whether, for this, it is necessary to observe his collection and its separation. If it were, the number o would be a puzzle; for up to now no one, I take it, has ever seen or touched o pebbles. MILL, of course, would explain o as something that has no sense, a mere manner of speaking; calculations with o would be a mere game, played with empty symbols, and the only wonder would be that anything rational could come of it. If, however, these calculations have a serious meaning, then the symbol o cannot be entirely without sense either. And the possibility suggests itself that $2 + 1$, in the same way as o, might have a sense even without MILL's matter of fact being observed. Who is actually prepared to assert that the fact which, according to MILL, is contained in the definition of an eighteen-figure number has ever been observed, and who is prepared to deny that the symbol for such a number has, none the less, a sense?

Perhaps it is supposed that the physical facts would be used only for the smaller numbers, say up to 10, while the remaining numbers could be constructed out of these. But if we can form 11 from 10 and 1 simply by definition, without having seen the corresponding collection, then there is no reason why we should not also be able in this way to construct 2 out of 1 and 1. If calculations with the number 11 do not follow from any matter of fact uniquely characteristic of that number, how does it happen that calculations with the number

2 must depend on the observation of a particular collection, separated in its own peculiar way?

It may, perhaps, be asked how arithmetic could exist, if we could distinguish nothing whatever by means of our senses, or only three things at most. Now for our knowledge, certainly, of arithmetical propositions and of their applications, such a state of affairs would be somewhat awkward—but would it affect the truth of those propositions? If we call a proposition empirical on the ground that we must have made observations in order to have become conscious of its content, then we are not using the word "empirical" in the sense in which it is opposed to "a priori". We are making a psychological statement, which concerns solely the content of the proposition; the question of its truth is not touched. In this sense, all Münchhausen's tales are empirical too; for certainly all sorts of observations must have been made before they could be invented.

Are the laws of arithmetic inductive truths?

§ 9. The considerations adduced thus far make it probable that numerical formulae can be derived from the definitions of the individual numbers alone by means of a few general laws, and that these definitions neither assert observed facts nor presuppose them for their legitimacy. Our next task, therefore, must be to ascertain the nature of the laws involved.

MILL[1] proposes to make use, for his proof (referred to above) of the formula $5 + 2 = 7$, of the principle that "Whatever is made up of parts, is made up of parts of those parts." This he holds to be an expression in more characteristic language of the principle familiar elsewhere in the form "The sums of equals are equals." He calls it an inductive truth, and a law of nature of the highest order. It is typical of the inaccuracy of

[1] Op. cit., Bk. III, cap. xxiv, § 5.

his exposition, that when he comes to the point in the proof at which, on his own view, this principle should be indispensable, he does not invoke it at all; however, it appears that his inductive truth is meant to do the work of LEIBNIZ's axiom that "If equals be substituted for equals, the equality remains." But in order to be able to call arithmetical truths laws of nature, MILL attributes to them a sense which they do not bear. For example,[1] he holds that the identity $1 = 1$ could be false, on the ground that one pound weight does not always weigh precisely the same as another. But the proposition $1 = 1$ is not intended in the least to state that it does.

MILL understands the symbol $+$ in such a way that it will serve to express the relation between the parts of a physical body or of a heap and the whole body or heap; but such is not the sense of that symbol. That if we pour 2 unit volumes of liquid into 5 unit volumes of liquid we shall have 7 unit volumes of liquid, is not the meaning of the proposition $5 + 2 = 7$, but an application of it, which only holds good provided that no alteration of the volume occurs as a result, say, of some chemical reaction. MILL always confuses the applications that can be made of an arithmetical proposition, which often are physical and do presuppose observed facts, with the pure mathematical proposition itself. The plus symbol can certainly look, in many applications, as though it corresponded to a process of heaping up; but that is not its meaning; for in other applications there can be no question of heaps or aggregates, or of the relationship between a physical body and its parts, as for example when we calculate about numbers of events. No doubt we can speak even here of "parts"; but then we are using the word not in the physical or geometrical sense, but in its logical sense, as we do when we speak of tyrannicides

[1] Op. cit., Bk. II, cap. vi, § 3.

as a part of murder as a whole. This is a matter of logical subordination. And in the same way addition too does not in general correspond to any physical relationship. It follows that the general laws of addition cannot, for their part, be laws of nature.

§ 10. But might they not still be inductive truths nevertheless? I do not see how that is conceivable. From what particular facts are we to take our start here, in order to advance to the general? The only available candidates for the part are the numerical formulae. Assign them to it, and of course we lose once again the advantage gained by giving our definitions of the individual numbers; we should have to cast around for some other means of establishing the numerical formulae. Even if we manage to rise superior to this misgiving too, which is not exactly easy, we shall still find the ground unfavourable for induction; for here there is none of that uniformity, which in other fields can give the method a high degree of reliability. LEIBNIZ[1] recognized this already: for to his Philalèthe, who had asserted that

"the several modes of number are not capable of any other difference but more or less; which is why they are simple modes, like those of space,"*

he returns the answer:

"That can be said of time and of the straight line, but certainly not of the figures and still less of the numbers, which are not merely different in magnitude, but also dissimilar. An even number can be divided into two equal parts, an odd number cannot; three and six are triangular numbers, four and nine are squares, eight is a cube, and so on. And this is even more the case with the numbers than with the figures; for two unequal figures can be perfectly similar to each other, but never two numbers."

We have no doubt grown used to treating the numbers

---

[1] Baumann, *Die Lehren von Raum, Zeit und Mathematik*, Vol. II, p. 39 (Erdmann edn., p. 243).

* [Derived from Locke, *Essay*, Bk. II, cap. xvi, § 5.]

in many contexts as all of the same sort, but that is only because we know a set of general propositions which hold for all numbers. For the present purpose, however, we must put ourselves in the position that none of these has yet been discovered. The fact is that it would be difficult to find an example of an inductive inference to parallel our present case. In ordinary inductions we often make good use of the proposition that every position in space and every moment in time is as good in itself as every other. Our results must hold good for any other place and any other time, provided only that the conditions are the same. But in the case of the numbers this does not apply, since they are not in space or time. Position in the number series is not a matter of indifference like position in space.

The numbers, moreover, are related to one another quite differently from the way in which the individual specimens of, say, a species of animal are. It is in their nature to be arranged in a fixed, definite order of precedence; and each one is formed in its own special way and has its own unique peculiarities, which are specially prominent in the cases of 0, 1 and 2. Elsewhere when we establish by induction a proposition about a species, we are ordinarily in possession already, merely from the definition of the concept of the species, of a whole series of its common properties. But with the numbers we have difficulty in finding even a single common property which has not actually to be first proved common.

The following is perhaps the case with which our putative induction might most easily be compared. Suppose we have noticed that in a borehole the temperature increases regularly with the depth; and suppose we have so far encountered a wide variety of differing rock strata. Here it is obvious that we cannot, simply on the strength of the observations made at this borehole, infer anything whatever as to the nature of the strata at deeper levels, and that any answer to the question, whether the regular distribution of temperature would continue to hold good lower down, would be premature. There is, it is true, a concept, that of "whatever you come to by going on boring," under which fall both the strata so far observed and those at lower levels alike; but that is of little assistance to us here. And equally, it will be no help to us to learn in the case of the numbers that these all fall together under the concept of "whatever you get by going on increasing by one." It is possible to draw a distinction between the two cases, on the ground that the strata are things we simply encounter, whereas the numbers are literally created, and determined in their whole natures, by the process of continually increasing by one. Now this can only mean that from the way in which a number, say 8, is generated through increasing by one all its properties can be deduced. But this is in principle to grant that the properties of numbers follow from their definitions, and to open up the possibility that we might prove the general laws of numbers from the method of generation which is common to them all, while deducing the special properties of the individual numbers from the special way in which, through the process of continually increasing by one, each one is formed. In the same way in the geological case too, we can deduce everything that is determined simply and solely by the depth at which a stratum is encountered, namely its spatial position relative to anything else, from the depth itself, without having any need of induction; but whatever is not so determined, cannot be learned by induction either.

The procedure of induction, we may surmise, can itself be justified only by means of general propositions of arithmetic —unless we understand by induction a mere process of habituation, in which case it has of course absolutely no power whatever of leading to the discovery of truth. The procedure of the sciences, with its objective standards, will at times find a high probability established by a single confirmatory instance, while at others it will dismiss a thousand as almost worthless; whereas our habits are determined by the number and strength of the impressions we receive and by subjective circumstances, which have no sort of right at all to influence our judgement. Induction [then, properly understood,] must base itself on the theory of probability, since it can never render a proposition more than probable. But how probability

theory could possibly be developed without presupposing
arithmetical laws is beyond comprehension.

§ 11. LEIBNIZ[1] holds the opposite view, that the
necessary truths, such as are found in arithmetic, must have
principles whose proof does not depend on examples and
therefore not on the evidence of the senses, though doubtless
without the senses it would have occurred to no one to think
of them. "The whole of arithmetic is innate and is in virtual
fashion in us." What he means by the term "innate" is
explained by another passage, where he denies "that *Everything
we learn is not innate*. The truths of number are in us and yet
we still learn them, whether it be by drawing them forth from
their source when learning them by demonstration (which
shows them to be innate), or whether it be . . .".

Are the laws of arithmetic synthetic a priori or analytic?

§ 12. If we now bring in the other antithesis between
analytic and synthetic, there result four possible combinations,
of which however one, viz.

analytic a posteriori

can be eliminated. Those who have decided with MILL in
favour of a posteriori have therefore no second choice, so that
there remain only two possibilities for us still to examine, viz.

synthetic a priori

and

analytic.

KANT declares for the former. In that case, there is no

---

[1] Baumann, op. cit., Vol. II, pp. 13-14 (Erdmann edn., pp. 195, 208-9).
[2] Baumann, op. cit., Vol. II, p. 38 (Erdmann edn., p. 212).

alternative but to invoke a pure intuition as the ultimate ground of our knowledge of such judgements, hard though it is to say of this whether it is spatial or temporal, or whatever else it may be. BAUMANN[1] agrees with KANT, although for rather different reasons. LIPSCHITZ,[2] too, holds that certain propositions, namely that which asserts that Number is independent of the method of numbering and also the Commutative and Associative Laws of Addition, are derived from inner intuition. HANKEL[3] bases the theory of real numbers on three fundamental propositions, to which he ascribes the character of "common notions" (*notiones communes*): "Once expounded they are perfectly self-evident; they are valid for magnitudes in every field, as vouched for by our pure intuition of magnitude; and they can without losing their character be transformed into definitions, simply by defining the addition of magnitudes as an operation which satisfies them." In the last statement here there is an obscurity. The definition can perhaps be constructed, but it will not do as a substitute for the original propositions; for in seeking to apply it the question would always arise: Are Numbers magnitudes, and is what we ordinarily call addition of Numbers addition in the sense of this definition? And to answer it, we should need to know already his original propositions about Numbers. Moreover, the expression "pure intuition of magnitude" gives us pause. If we consider all the different things that are called magnitudes: Numbers, lengths, areas, volumes, angles, curvatures, masses, velocities, forces, illuminations, electric currents, and so forth, we can quite well understand how they can all be brought under the single *concept* of magnitude; but the term "intuition of magnitude," and still worse "pure intuition of

---

[1] Op. cit., Vol. II, p. 669.
[2] *Lehrbuch der Analysis*, Vol. I, p. 1 [Bonn 1877].
[3] *Theorie der complexen Zahlensysteme*, pp. 54–55.

magnitude", cannot be admitted as appropriate. I cannot even allow an intuition of 100,000, far less of number in general, not to mention magnitude in general. We are all too ready to invoke inner intuition, whenever we cannot produce any other ground of knowledge. But we have no business, in doing so, to lose sight altogether of the sense of the word "intuition".

KANT in his *Logic* (ed. Hartenstein, vol. VIII, p. 88) defines it as follows:

"An intuition is an *individual* idea (REPRÆSENTATIO SINGULARIS), a concept is a *general* idea (REPRÆSENTATIO PER NOTAS COMMUNES) or an idea *of reflexion* (REPRÆSENTATIO DISCURSIVA)."

Here there is absolutely no mention of any connexion with sensibility, which is, however, included in the notion of intuition in the *Transcendental Aesthetic*, and without which intuition cannot serve as the principle of our knowledge of synthetic a priori judgements. In the *Critique of Pure Reason* (ed. Hartenstein, vol. III, p. 55)* we read:

"It is therefore through the medium of sensibility that objects are *given* to us and it alone provides us with *intuitions*."

It follows that the sense of the word "intuition" is wider in the *Logic* than in the *Transcendental Aesthetic*. In the sense of the *Logic*, we might perhaps be able to call 100,000 an intuition; for it is not a general concept anyhow. But an intuition in this sense cannot serve as the ground of our knowledge of the laws of arithmetic.

§ 13. We shall do well in general not to overestimate the extent to which arithmetic is akin to geometry. I have already quoted a warning to this effect from LEIBNIZ. One geometrical point, considered by itself, cannot be distinguished in any way from any other; the same applies to lines and planes. Only when several points, or lines or planes, are included together in a single intuition, do we distinguish them. In geometry, therefore, it is quite intelligible that general pro-

---

* [Original edns., A19/B33]

positions should be derived from intuition; the points or lines or planes which we intuite are not really particular at all, which is what enables them to stand as representatives of the whole of their kind. But with the numbers it is different; each number has its own peculiarities. To what extent a given particular number can represent all the others, and at what point its own special character comes into play, cannot be laid down generally in advance.

§ 14. If, again, we compare the various kinds of truths in respect of the domains that they govern, the comparison tells once more against the supposed empirical and synthetic character of arithmetical laws.

Empirical propositions hold good of what is physically or psychologically actual, the truths of geometry govern all that is spatially intuitable, whether actual or product of our fancy. The wildest visions of delirium, the boldest inventions of legend and poetry, where animals speak and stars stand still, where men are turned to stone and trees turn into men, where the drowning haul themselves up out of swamps by their own topknots—all these remain, so long as they remain intuitable, still subject to the axioms of geometry. Conceptual thought alone can after a fashion shake off this yoke, when it assumes, say, a space of four dimensions or positive curvature. To study such conceptions is not useless by any means; but it is to leave the ground of intuition entirely behind. If we do make use of intuition even here, as an aid, it is still the same old intuition of Euclidean space, the only one whose structures we can intuit. Only then the intuition is not taken at its face value, but as symbolic of something else; for example, we call straight or plane what we actually intuite as curved. For purposes of conceptual thought we can always assume the contrary of some one or other of the geometrical axioms, without involving ourselves in any self-contradictions when we proceed to our deductions, despite the conflict between our assumptions and our intuition. The fact that this is possible shows that the axioms of geometry are independent of one another and of the primitive laws of logic, and consequently are synthetic. Can the same be said of the fundamental propositions of the science of number? Here, we have only to try denying any one of them, and complete confusion ensues. Even to think at all seems no longer possible. The basis of arithmetic lies deeper, it seems, than that of any of the empirical sciences, and even than that of geometry. The truths of arithmetic govern all that is numerable. This is the widest domain of all; for to it belongs not only the actual, not only the intuitable, but everything thinkable. Should not the laws of number, then, be connected very intimately with the laws of thought?

§ 15. Statements in LEIBNIZ can only be taken to mean that the laws of number are analytic, as was to be expected, since for him the a priori coincides with the analytic. Thus he declares[1] that the benefits of algebra are due to its borrowings from a far superior science, that of the true logic. In another passage[2] he compares necessary and contingent truths to commensurable and incommensurable magnitudes, and maintains that in the case of necessary truths a proof or reduction to identities* is possible. However, these declarations lose some of their force in view of LEIBNIZ's[3] inclination to regard all truths as provable: "Every truth", he says, "has its proof a priori derived from the concept of the terms, notwithstanding it does not always lie in our power to achieve this analysis." Though of course the comparison to commensurable and incommensurable magnitudes erects a fresh

[1] Baumann, op. cit., Vol. II, p. 56 (Erdmann edn., p. 424).
[2] Baumann, op. cit., Vol. II, p. 57 (Erdmann edn., p. 83).
[3] Baumann, op. cit., Vol. II, p. 57 (Pertz edn., Vol. II, p. 55 [=Gerhardt edn., phil. Schr., Vol. II, p. 62]).
* [Identitäten]

barrier between necessary and contingent truths, which for us at least is insuperable.

A very emphatic declaration in favour of the analytic nature of the laws of number is that of W. S. JEVONS[1]: "I hold that algebra is a highly developed logic, and number but logical discrimination."

§ 16. But this view, too, has its difficulties. Can the great tree of the science of number as we know it, towering, spreading, and still continually growing, have its roots in bare identities*? And how do the empty forms of logic come to disgorge so rich a content?

To quote MILL:[2] "The doctrine that we can discover facts, detect the hidden processes of nature, by an artful manipulation of language, is so contrary to common sense, that a person must have made some advances in philosophy to believe it."

Very true—if it be supposed that during the artful manipulation we do not think at all. MILL is here criticizing a kind of formalism that scarcely anyone would wish to defend. Everyone who uses words or mathematical symbols makes the claim that they mean something, and no one will expect any sense to emerge from empty symbols. But it is possible for a mathematician to perform quite lengthy calculations without understanding by his symbols anything intuitable, or with which we could be sensibly acquainted. And that does not mean that the symbols have no sense; we still distinguish between the symbols themselves and their content, even though it may be that the content can only be grasped by their aid. We realize perfectly that other symbols might have been assigned to stand for the same things. All we need to know is how to handle logically the content as made sensible in the symbols and, if we wish to apply our calculus to physics, how to effect the transition to the phenomena.

[1] *The Principles of Science*, London 1879, p. 156 [1874 edn., p. 174].
[2] Op. cit., Bk. II, cap. vi, § 2.

* [*Identitäten*]

It is, however, a mistake to see in such applications the real sense of the propositions; in any application a large part of their generality is always lost, and a particular element enters in, which in other applications is replaced by other particular elements.

§ 17. However much we may disparage deduction, it cannot be denied that the laws established by induction are not enough. New propositions must be derived from them which are not contained in any one of them by itself. No doubt these propositions are in a way contained covertly in the whole set taken together, but this does not absolve us from the labour of actually extracting them and setting them out in their own right. This seen, we can see also the following possibility. Instead of linking our chain of deductions direct to any matter of fact, we can leave the fact on one side, while adopting its content in the form of a condition. By substituting in this way conditions for facts throughout the whole of a train of reasoning, we shall finally reduce it to a form in which a certain result is made dependent on a certain series of conditions. This truth would be established by thought alone or, to use MILL's expression, by an artful manipulation of language. It is not impossible that the laws of number are of this type. This would make them analytic judgements, despite the fact that they would not normally be discovered by thought alone; for we are concerned here not with the way in which they are discovered but with the kind of ground on which their proof rests; or in LEIBNIZ's[1] words, "the question here is not one of the history of our discoveries, which is different in different men, but of the connexion and natural order of truths, which is always the same." It would then rest with observation finally to decide whether the conditions included in the laws thus established are actually fulfilled. Thus we should in the end arrive at the same position as we should have reached by linking our chain

[1] *Nouveaux Essais*, IV, § 9 (Erdmann edn., p. 362).

of deductions direct to observed matters of fact. But the type of procedure here indicated is in many cases to be preferred, because it leads to a general proposition, which need not be applicable only to the facts immediately under consideration. The truths of arithmetic would then be related to those of logic in much the same way as the theorems of geometry to the axioms. Each one would contain concentrated within it a whole series of deductions for future use, and the use of it would be that we need no longer make the deductions one by one, but can express simultaneously the result of the whole series.[1] If this be so, then indeed the prodigious development of arithmetical studies, with their multitudinous applications, will suffice to put an end to the widespread contempt for analytic judgements and to the legend of the sterility of pure logic.

This is not the first time that such a view has been put forward. If it could be worked out in detail, so rigorously that not the smallest doubt remained, that, it seems to me, would be a result not entirely without importance.

II. Views of certain writers on the concept of Number.

§ 18. On turning now to consider the primary objects of arithmetic, we must distinguish between the individual numbers 3, 4 and so on, and the general concept of Number.

---

[1] It is remarkable that MILL too (op. cit., Bk. II, cap. vi, § 4) seems to express this view. His sound sense, in fact, from time to time breaks through his prejudice in favour of the empirical. But this same prejudice as often brings everything back into a muddle, by making him confuse the physical applications of arithmetic with arithmetic itself. He seems to be unaware that a hypothetical judgement can be true even when the antecedent is not true.

Now we have already decided in favour of the view that the individual numbers are best derived, in the way proposed by LEIBNIZ, MILL, H. GRASSMANN and others, from the number one together with increase by one, but that these definitions remain incomplete so long as the number one and increase by one are themselves undefined. And we have seen that we have need of general propositions if we are to derive the numerical formulae from these definitions. Such laws cannot, just because of their generality, follow from the definitions of the individual numbers, but only from the general concept of Number. It is this concept that we shall now submit to a closer examination; in the course of this we may expect to have also to discuss the number one and increase by one, as a result of which in turn we shall expect to complete the definitions of the individual numbers.

§ 19. At this point I should like straight away to oppose the attempt to think of number geometrically, as a ratio between lengths or surfaces. Obviously, the thought behind this was to facilitate the numerous applications of arithmetic to geometry by putting the rudiments of both in the closest connexion from the outset.

NEWTON[1] proposes to understand by number not so much a set of units as the relation in the abstract between any given magnitude* and another magnitude of the same kind which is taken as unity. It may be granted that this is an apt description of number in the wider sense, in which it includes [besides the integers] also fractions and irrational numbers; but it presupposes the concepts of magnitude and of relation in respect of magnitude. This should presumably mean that a definition of number in the narrower sense, or cardinal Number, will still be needed; for EUCLID**, in order to define the identity of two ratios between lengths, makes use of the concept of equimultiples, and equimultiples bring us back once again

---

[1] Baumann, op. cit., Vol. I, p. 475 [*Arithmetica Universalis*, Vol. I, cap. ii, 3.]

* [*quantitas*]

** [*Elements*, Bk. V., Def. 5.]

to numerical identity. However, let it be, as it may be, the case that identity of ratios between lengths can in fact be defined without any reference to the concept of number. Even so, we should still remain in doubt as to how the number defined geometrically in this way is related to the number of ordinary life, which would then be entirely cut off from science. Yet surely we are entitled to demand of arithmetic that its numbers should be adapted for use in every application made of number, even although that application is not itself the business of arithmetic. Even in our everyday sums, we must be able to rely on the science of arithmetic to provide the basis for the methods we use. And moreover, the question arises whether arithmetic itself can make do with a geometrical concept of number, when we think of some of the notions that occur in it, such as the Number of roots of an equation or of numbers prime to and smaller than a given number. On the other hand, the number which gives the answer to the question *How many?* can answer among other things how many units are contained in a length. And operations with negative, fractional and irrational numbers can all be reduced to operations with the natural numbers. Perhaps what NEWTON wished to understand by magnitudes, in defining number as a relation between magnitudes, was not geometrical magnitudes only, but also sets. In that case, however, his definition is useless for our purposes, since the expression "relation between a set and the unit of the set" tells us no more than the expression "number by which a set is determined."

§ 20. The first question to be faced, then, is whether number is definable. HANKEL[1] declares that it is not, in these words: "What we mean by thinking or putting a thing once, twice, three times, and so on, cannot be defined, because of the simplicity in principle of the concept of putting." But the point is surely not so much the putting as the once, twice and three times. If this could be defined, the indefinability

---

[1] Op. cit., p. 1.

of putting would scarcely worry us. LEIBNIZ is inclined to regard number as an adequate idea, meaning one which is so clear that every element contained in it is also clear, or at least as an almost adequate one. If the general inclination is, on the whole, to hold that Number is indefinable, that is more because attempts to define it have failed than because anything has been discovered in the nature of the case to show that it must be so.

Is Number a property of external things?

§ 21. Let us try at least to assign to Number its proper place among our concepts. In language, numbers most commonly appear in adjectival form and attributive construction in the same sort of way as the words hard or heavy or red, which have for their meanings properties of external things. It is natural to ask whether we must think of the individual numbers too as such properties, and whether, accordingly, the concept of Number can be classed along with that, say, of colour.

That it can, seems to be the view of M. CANTOR,[1] when he calls mathematics an empirical science in so far as it begins with the consideration of things in the external world. On his view, number originates only by abstraction from objects.

For E. SCHRÖDER[2] number is modelled on actuality, derived from it by a process of copying the actual units with ones, which he calls the abstraction of number. In this copying, the units are only represented in point of their frequency, all

---

[1] *Grundzüge einer Elementarmathematik*, p. 2, § 4. Similarly Lipschitz, op. cit., p. 1.

[2] Op. cit., pp. 6, 10–11.

other properties of the things concerned, such as their colour or shape, being disregarded. Here frequency is only another name for Number. It follows, therefore, that SCHRÖDER puts frequency or Number on a level with colour and shape, and treats it as a property of things.

§ 22. BAUMANN[1] rejects the view that numbers are concepts extracted from external things: "The reason being that external things do not present us with any strict units; they present us with isolated groups or sensible points, but we are at liberty to treat each one of these itself again as a many." And it is quite true that, while I am not in a position, simply by thinking of it differently, to alter the colour or hardness of a thing in the slightest, I am able to think of the Iliad either as one poem, or as 24 Books, or as some large Number of verses. Is it not in totally different senses that we speak of a tree as having 1000 leaves and again as having green leaves? The green colour we ascribe to each single leaf, but not the number 1000. If we call all the leaves of a tree taken together its foliage, then the foliage too is green, but it is not 1000. To what then does the property 1000 really belong? It almost looks as though it belongs neither to any single one of the leaves nor to the totality of them all; is it possible that it does not really belong to things in the external world at all? If I give someone a stone with the words: Find the weight of this, I have given him precisely the object he is to investigate. But if I place a pile of playing cards in his hands with the words: Find the Number of these, this does not tell him whether I wish to know the number of cards, or of complete packs of cards, or even say of points in the game of skat. To have given him the pile in his hands is not yet to have given him completely the object he is to investigate; I must add some

---

[1] Op. cit., Vol. II, p. 669.

further word—cards, or packs, or points. Nor can we say that in this case the different numbers exist in the same thing side by side, as different colours do. I can point to the patch of each individual colour without saying a word, but I cannot in the same way point to the individual numbers. If I can call the same object red and green with equal right, it is a sure sign that the object named is not what really has the green colour; for that we must first get a surface which is green only. Similarly, an object to which I can ascribe different numbers with equal right is not what really has a number.

It marks, therefore, an important difference between colour and Number, that a colour such as blue belongs to a surface independently of any choice of ours. The blue colour is a power of reflecting light of certain wavelengths and of absorbing to varying extents light of other wavelengths; to this, our way of regarding it cannot make the slightest difference. The Number 1, on the other hand, or 100 or any other Number, cannot be said to belong to the pile of playing cards in its own right, but at most to belong to it in view of the way in which we have chosen to regard it; and even then not in such a way that we can simply assign the Number to it as a predicate. What we choose to call a complete pack is obviously an arbitrary decision, in which the pile of playing cards has no say. But it is when we examine the pile in the light of this decision, that we discover perhaps that we can call it two complete packs. Anyone who did not know what we call a complete pack would probably discover in the pile any other Number you like before hitting on two.

§ 23. To the question: What is it that the number belongs to as a property? MILL[1] replies as follows: the name of a number connotes, "of course, some property belonging to the

---

[1] Op. cit., Bk. III, cap. xxiv, § 5.

agglomeration of things which we call by the name; and that property is the characteristic manner in which the agglomeration is made up of, and may be separated into, parts."

Here the definite article in the phrase "the characteristic manner" is a mistake right away; for there are very various manners in which an agglomeration can be separated into parts, and we cannot say that one alone would be characteristic. For example, a bundle of straw can be separated into parts by cutting all the straws in half, or by splitting it up into single straws, or by dividing it into two bundles. Further, is a heap of a hundred grains of sand made up of parts in exactly the same way as a bundle of 100 straws? And yet we have the same number. The number word "one", again, in the expression "one straw" signally fails to do justice to the way in which the straw is made up of cells or molecules. Still more difficulty is presented by the number o. Besides, need the straws form any sort of bundle at all in order to be numbered? Must we literally hold a rally of all the blind in Germany before we can attach any sense to the expression "the number of blind in Germany"? Are a thousand grains of wheat, when once they have been scattered by the sower, a thousand grains of wheat no longer? Do such things really exist as agglomerations of proofs of a theorem, or agglomerations of events? And yet these too can be numbered. Nor does it make any difference whether the events occur together or thousands of years apart.

§ 24. This brings us to another reason for refusing to class number along with colour and solidity: it is applicable over a far wider range.

MILL[1] maintains that the truth that whatever is made up of parts is made up of parts of those parts holds good for natural phenomena of every sort, since all admit of being

---

[1] Op. cit., Bk. III, cap. xxiv, § 5.

numbered. But cannot still far more than this be numbered?
LOCKE[1] says: "Number applies itself to men, angels, actions,
thoughts—everything that either doth exist or can be
imagined." LEIBNIZ[2] rejects the view of the schoolmen that
number is not applicable to immaterial things, and calls
number a sort of immaterial figure, which results from the
union of things of any sorts whatsoever, for example of God,
an angel, a man and motion, which together are four. For
which reason he holds that number is of supreme universality
and belongs to metaphysics. In another passage[3] he says:
"Some things cannot be weighed, as having no force and
power; some things cannot be measured, by reason of having
no parts; but there is nothing which cannot be numbered.
Thus number is, as it were, a kind of metaphysical figure."

It would indeed be remarkable if a property abstracted
from external things could be transferred without any change
of sense to events, to ideas and to concepts. The effect would
be just like speaking of fusible events, or blue ideas, or salty
concepts or tough judgements.

It does not make sense that what is by nature sensible
should occur in what is non-sensible. When we see a blue
surface, we have an impression of a unique sort, which
corresponds to the word "blue"; this impression we recognize
again, when we catch sight of another blue surface. In order
to suppose that there is in the same way, when we look at a
triangle, something sensible corresponding to the word
"three", we should have to commit ourselves to finding that
same thing again in three concepts too; so that something
non-sensible would have in it something sensible. It may

---

[1] Baumann, op. cit., Vol. I, p. 409. [*Essay*, Bk. II, cap. xvi, § 1].
[2] Baumann, op. cit., Vol. II, pp. 2-3 [Erdmann edn., p. 8].
[3] Baumann, op. cit., Vol. II, p. 56 [Erdmann edn., p. 162].

certainly be granted that a sensible impression of a sort does correspond to the word "triangular", but then the word must be taken as a whole. The three in it we do not see directly; rather, we see something upon which can fasten an intellectual activity of ours leading to a judgement in which the number 3 occurs. How is it after all that we do become acquainted with, let us say, the Number of figures of the syllogism as drawn up by Aristotle? Is it perhaps with our eyes? What we see is at most certain symbols for the syllogistic figures, not the figures themselves. How are we to be able to see their Number, if they themselves remain invisible? However, it may be argued that it is enough to see the symbols; their number is identical with the number of the figures. But then, how do we know this? For that, we must have already ascertained the number of the figures by some other means. Or is the proposition "The Number of figures of the syllogism is four" only another way of putting the proposition that "The Number of symbols for figures of the syllogism is four?" Of course it is not. There is no intention of saying anything about the symbols; no one wants to know anything about them, except in so far as some property of theirs directly mirrors some property in what they symbolize. Besides, the same thing can, without any logical fallacy, be symbolized by several different symbols, so that there is not even any need for the number of symbols to coincide with the number of things symbolised.

§ 25. While for MILL the number is something physical, for LOCKE and LEIBNIZ it exists only as a notion.* MILL[1] is, of course, quite right that two apples are physically different from three apples, and two horses from one horse; that they are a different visible and tangible phenomenon.[2] But are

---

[1] Op. cit., Bk. III, cap. xxiv, § 5.

[2] Strictly speaking we should add: provided that they are a phenomenon at all. For if someone has one horse in Germany and one in America (and no others), then he does possess two horses; yet these two horses do not form a phenomenon,—only each one of the two by itself could be so described.

* [in der Idee]

we to infer from this that their twoness or threeness is something physical? *One* pair of boots may be the same visible and tangible phenomenon as *two* boots. Here we have a difference in number to which no physical difference corresponds; for *two* and *one pair* are by no means the same thing, as MILL seems oddly to believe. How is it possible, after all, for two concepts to be physically distinguishable from three concepts?

To quote BERKELEY[1]: "It ought to be considered that number . . . is nothing fixed and settled, really existing in things themselves. It is entirely the creature of the mind, considering, either an idea by itself, or any combination of ideas to which it gives one name, and so makes it pass for a unit. According as the mind variously combines its ideas, the unit varies; and as the unit, so the number, which is only a collection of units, doth also vary. We call a window one, a chimney one, and yet a house in which there are many windows, and many chimneys, hath an equal right to be called one, and many houses go to the making of one city."

Is number something subjective?

§ 26. This line of thought may easily lead us to regard number as something subjective. It looks as though the way in which number originates in us may prove the key to its essential nature. The matter would thus become one for a psychological enquiry. This is indeed what LIPSCHITZ[2] is thinking of when he writes: "Anyone who proposes to make a survey of a number of things, will begin with some one particular thing and proceed by continually adding a new one to those previously selected." This seems to describe much better how we acquire say the intuition of a constellation than how we construct numbers. The intention to make a

---

[1] Baumann, op. cit., Vol. II, p. 428 [*New Theory of Vision*, § 109].
[2] Op. cit., p. 1. I take it that Lipschitz means to refer to a mental process.

survey is not essential; for it will scarcely be maintained that it becomes any easier to survey a herd after we have learned how many head it comprises.

No description of this kind of the mental processes which precede the forming of a judgement of number*, even if more to the point than this one, can ever take the place of a genuine definition of the concept. It can never be adduced in proof of any proposition of arithmetic; it acquaints us with none of the properties of numbers. For number is no whit more an object of psychology or a product of mental processes than, let us say, the North Sea is. The objectivity of the North Sea is not affected by the fact that it is a matter of our arbitrary choice which part of all the water on the earth's surface we mark off and elect to call the "North Sea". This is no reason for deciding to investigate the North Sea by psychological methods. In the same way number, too, is something objective. If we say "The North Sea is 10,000 square miles in extent" then neither by "North Sea" nor by "10,000" do we refer to any state of or process in our minds: on the contrary, we assert something quite objective, which is independent of our ideas and everything of the sort. If we should happen to wish, on another occasion, to draw the boundaries of the North Sea differently or to understand something different by "10,000", that would not make false the same content that was previously true: what we should perhaps rather say is, that a false content had now taken the place of a true, without in any way depriving its predecessor of its truth.

The botanist means to assert something just as factual when he gives the Number of a flower's petals as when he gives their colour. The one depends on our arbitrary choice just as little as the other. There does, therefore, exist a certain similarity between Number and colour; it consists, however, not in our becoming acquainted with them both in external things through the senses, but in their being both objective.

* [For Frege, a "judgement of number" (*Zahlurtheil*), like its verbal expression, a "statement of number" (*Zahlangabe*), is one as to *how many* of something there are.]

I distinguish what I call objective from what is handleable or spatial or actual. The axis of the earth is objective, so is the centre of mass of the solar system, but I should not call them actual in the way the earth itself is so. We often speak of the equator as an *imaginary* line; but it would be wrong to call it a *fictitious* line; it is not a creature of thought, the product of a psychological process, but is only recognized or apprehended by thought. If to be recognized were to be created, then we should be able to say nothing positive about the equator for any period earlier than the date of its alleged creation.

Space, according to KANT, belongs to appearance. For other rational beings it might take some form quite different from that in which we know it. Indeed, we cannot even know whether it appears the same to one man as to another; for we cannot, in order to compare them, lay one man's intuition of space beside another's. Yet there is something objective in it all the same; everyone recognizes the same geometrical axioms, even if only by his behaviour, and must do so if he is to find his way about the world. What is objective in it is what is subject to laws, what can be conceived and judged, what is expressible in words. What is purely intuitable is not communicable. To make this clear, let us suppose two rational beings such that projective properties and relations are all they can intuite—the lying of three points on a line, of four points on a plane, and so on; and let what the one intuites as a plane appear to the other as a point, and vice versa, so that what for the one is the line joining two points for the other is the line of intersection of two planes, and so on with the one intuition always dual to the other. In these circumstances they could understand one another quite well and would never realize the difference between their intuitions, since in projective geometry every proposition has its dual counterpart; any disagreements over points of aesthetic appreciation would not be conclusive evidence. Over all geometrical theorems they would be in complete agreement, only interpreting the words differently in terms of their respective intuitions. With the word "point", for example, one would connect one intuition and the other another. We can therefore still say that this word has for them an objective meaning, provided only that by this meaning we do not understand any of the peculiarities of their respective intuitions. And in this sense the axis of the earth too is objective.

The word "white" ordinarily makes us think of a certain sensation, which is, of course, entirely subjective; but even in ordinary everyday speech, it often bears, I think, an objective sense. When we call snow white, we mean to refer to an objective quality which we recognize, in ordinary daylight, by a certain sensation. If the snow is being seen in a coloured light, we take that into account in our judgement and say, for instance, "It *appears* red at present, but it *is* white." Even a colour-blind man can speak of red and green, in spite of the fact that he does not distinguish between these colours in his sensations; he recognizes the distinction by the fact that others make it, or perhaps by making a physical experiment. Often, therefore, a colour word does not signify our subjective sensation, which we cannot know to agree with anyone else's (for obviously our calling things by the same name does not guarantee as much), but rather an objective quality. It is in this way that I understand objective to mean what is independent of our sensation, intuition and imagination, and of all construction of mental pictures out of memories of earlier sensations, but not what is independent of the reason,—for what are things independent of the reason? To answer that would be as much as to judge without judging, or to wash the fur without wetting it.

§ 27. For that reason I cannot agree with SCHLOEMILCH[1]

---

[1] *Handbuch der algebraischen Analysis*, p. 1.

either, when he calls number the idea of the position of an item in a series.[1] If number were an idea, then arithmetic would be psychology. But arithmetic is no more psychology than, say, astronomy is. Astronomy is concerned, not with ideas of the planets, but with the planets themselves, and by the same token the objects of arithmetic are not ideas either. If the number two were an idea, then it would have straight away to be private to me only. Another man's idea is, *ex vi termini*, another idea. We should then have it might be many millions of twos on our hands. We should have to speak of my two and your two, of one two and all twos. If we accept latent or unconscious ideas, we should have unconscious twos among them, which would then return subsequently to consciousness. As new generations of children grew up, new generations of twos would continually be being born, and in the course of millennia these might evolve, for all we could tell, to such a pitch that two of them would make five. Yet, in spite of all this, it

---

[1] Another possible objection is, that on this theory the same idea of a position in a series would have always to appear whenever the same number occurred, which obviously does not happen. My arguments would be beside the point if he meant by idea an objective notion [*Idee*]; but in that case what distinction would there be between the idea of the position and the position itself?

An idea in the subjective sense is what is governed by the psychological laws of association; it is of a sensible, pictorial character. An idea in the objective sense belongs to logic and is in principle non-sensible, although the word which means an objective idea is often accompanied by a subjective idea, which nevertheless is not its meaning. Subjective ideas are often demonstrably different in different men, objective ideas are the same for all. Objective ideas can be divided into objects and concepts. I shall myself, to avoid confusion, use "idea" only in the subjective sense. It is because Kant associated both meanings with the word that his doctrine assumed such a very subjective, idealist complexion, and his true view was made so difficult to discover. The distinction here drawn stands or falls with that between psychology and logic. If only these themselves were to be kept always rigidly distinct!

would still be doubtful whether there existed infinitely many numbers, as we ordinarily suppose. 10^{10}, perhaps, might be only an empty symbol, and there might exist no idea at all, in any being whatever, to answer to the name.

Weird and wonderful, as we see, are the results of taking seriously the suggestion that number is an idea. And we are driven to the conclusion that number is neither spatial and physical, like Mill's piles of pebbles and gingersnaps, nor yet subjective like ideas, but non-sensible and objective. Now objectivity cannot, of course, be based on any sense-impression, which as an affection of our mind is entirely subjective, but only, so far as I can see, on the reason.

It would be strange if the most exact of all the sciences had to seek support from psychology, which is still feeling its way none too surely.

Numbers as sets.

§ 28. Some writers define Number as a set or multitude or plurality. All these views suffer from the drawback that the concept will not then cover the numbers o and 1. Moreover, these terms are utterly vague: sometimes they approximate in meaning to "heap" or "group" or "agglomeration", referring to a juxtaposition in space, sometimes they are so used as to be practically equivalent to "Number", only vaguer. No analysis of the concept of Number, therefore, is to be found in a definition of this kind. THOMAE[1] requires for the formation of number that item-sets which differ be given different names. By this he evidently means to refer to a process of bringing out more sharply the characteristics of the sets in question, of which the giving of names is only the external sign. The question is, just what is this process like?

[1] *Elementare Theorie der analytischen Functionen*, p. 1.

Obviously, the notion of number would not result if, instead of "3 stars", "3 fingers" and "7 stars", we tried introducing names in which there were no recognizable common elements. It is not a matter simply of assigning names, but of symbolizing in its own right the numerical element. For this, we must needs have come to recognize that element in its peculiarity.

Furthermore, it should be noted that there are two different views. Some call number a set of things or objects; others, following EUCLID[1], define it as a set of units. This last expression demands a separate discussion.

III. Views on unity and one*.

Does the number word "one" stand for a property of objects?

§ 29. In the definitions which EUCLID gives at the beginning of Book VII of the Elements, he seems to mean by the word "μονάς" sometimes an object to be counted, sometimes a property of such an object, and sometimes the number one. We can translate it consistently by the German "Einheit", but only because that word itself shifts over the same variety of meanings.

According to SCHRÖDER[2]: "Each of the things to be counted is called a unit." We may well wonder why we must first bring the things under the concept of unity, instead of simply defining number right away as a set of things, which would throw us back once again onto the first of the two views. The most obvious answer is that in calling the things units we are supposed to be adding to our description of them; under the influence of the grammatical form, we are regarding "one"

---

[1] Μονάς ἐστι, καθ᾽ ἣν ἕκαστον τῶν ὄντων ἓν λέγεται. Ἀριθμὸς δὲ τὸ ἐκ μονάδων συγκείμενον πλῆθος. ["A unit is that by virtue of which each of the things that exist is called one. A number is a multitude composed of units."]

[2] Op. cit., p. 5.

* [It is not possible in English to do entire justice to the ambiguities of the German *Einheit*, which covers both "unit" and "unity", not to mention "oneness". Moreover *Einheit* is a verbal derivative of *Ein* (whereas "unit/y" is not derived directly from "one"), and derivatives of either word can be described alike as derivatives of *Ein* or "one". These facts make §§ 29, 32 and 37, in particular, more plausible in German than in English.]

as a word for a property and taking "one city" in the same way as "wise man". In that case a unit would be an object characterized by the property "one" and would stand to "one" in the same relation as "a sage" to the adjective "wise". Now reasons have already been given as conclusive against the view that number is a property of things; but there are several further arguments against the present suggestion in particular. It must strike us immediately as remarkable that every single thing should possess this property. It would be incomprehensible why we should still ascribe it expressly to a thing at all. It is only in virtue of the possibility of something not being wise that it makes sense to say "Solon is wise." The content of a concept diminishes as its extension increases; if its extension becomes all-embracing, its content must vanish altogether. It is not easy to imagine how language could have come to invent a word for a property which could not be of the slightest use for adding to the description of any object whatsoever.

If it were correct to take "one man" in the same way as "wise man", we should expect to be able to use "one" also as a grammatical predicate, and to be able to say "Solon was one" just as much as "Solon was wise". It is true that "Solon was one" can actually occur, but not in a way to make it intelligible on its own in isolation. It may, for example, mean "Solon was a wise man", if "wise man" can be supplied from the context. In isolation, however, it seems that "one" cannot be a predicate.[1] This is even clearer if we take the plural. Whereas we can combine "Solon was wise" and "Thales was wise" into "Solon and Thales were wise", we cannot say "Solon and Thales were one". But it is hard to see why this

---

[1] Usages do occur which appear to contradict this; but if we look more closely we shall find that some general term has to be supplied, or else that "one" is not being used as a number word—that what it is intended to assert is the character (not of being unique, but) of being unitary.

should be impossible, if "one" were a property both of Solon and of Thales in the same way that "wise" is.

§ 30. In line with this is the fact that no one has ever been able to give a definition of the property "one". LEIBNIZ[1] indeed says that "By *one* is meant whatever we grasp in one act of the understanding," but this is to define "one" in terms of itself. Besides, surely we can also grasp what is many in one act of the understanding? LEIBNIZ admits as much in the same passage. BAUMANN[2] does no better when he says: "That is one, which we apprehend as one", and further: "Whatever we take as a point, or refuse to take as further subdivided into parts, that we regard as one; but every one of outer intuition, whether empirical or pure, can also be regarded as a many. Every idea is one when isolated in contrast with another; but in itself it can again be distinguished into a many." This sweeps away every limit to the application of the concept imposed by the nature of the facts, and everything is made dependent on our way of regarding them. I ask once more: How can it make sense to ascribe the property "one" to any object whatever, when every object, according as to how we look at it, can be either one or not one? How can a science which bases its claim to fame precisely on being as definite and accurate as possible repose on a concept as hazy as this is?

§ 31. Now although BAUMANN[3] bases the concept of one on inner intuition, he refers nevertheless, in the passage just cited, to certain criteria for being one, namely being undivided and being isolated. If this were correct, then we should have to expect animals, too, to be capable of having some sort of idea of unity. Can it be that a dog staring at the moon does have an idea, however ill-defined, of what we signify by the

[1] Baumann, op. cit., Vol. II, p. 2 (Erdmann edn., p. 8).
[2] Op. cit., Vol. II, p. 669.
[3] Op. cit., Vol. II, p. 669.

word "one"? This is hardly credible—and yet it certainly distinguishes individual objects: another dog, its master, a stone it is playing with, these certainly appear to the dog every bit as isolated, as self-contained, as undivided, as they do to us. It will notice a difference, no doubt, between being set on by several other dogs and being set on by only one, but this is what MILL calls the physical difference. We need to know specifically: is the dog conscious, however dimly, of that common element in the two situations which we express by the word "one", when, for example, it first is bitten by one larger dog and then chases one cat? This seems to me unlikely. I infer, therefore, that the notion of unity is not, as LOCKE[1] holds, "suggested to the understanding by every object without us, and every idea within", but becomes known to us through the exercise of those higher intellectual powers which distinguish men from brutes. Consequently, such properties of things as being undivided or being isolated, which animals perceive quite as well as we do, cannot be what is essential in our concept.

§ 32. Still, we may suspect that they have some sort of connexion with it. Language indicates as much by forming "united" as a derivative from "one". The more the internal contrasts within a thing fade into insignificance by comparison with the contrasts between it and its environment, and the more the internal connexions among its elements overshadow its connexions with its environment, the more natural it becomes for us to regard it as a distinct object. For a thing to be "united" means that it has a property which causes us, when we think of it, to sever it from its environment and consider it on its own. In the same way we can explain how the French "uni" comes to mean "even" or "smooth". The word "unity" too, is used in a similar manner, when we speak of the political

---

[1] Baumann, op. cit., Vol. I, p. 409 [*Essay*, Bk. II, cap. vii, § 7].

unity of a country or the unity of a work of art.[1] But in this sense "unity" is connected not so much with "one" as with "united" or "unitary". For when we say the Earth has one moon, we do not mean to point out that our satellite is isolated, or self-contained, or undivided; rather, we are contrasting the satellite system of the Earth with that of Venus or Mars or Jupiter. So far as being isolated goes, or being undivided, the moons of Jupiter could stand up quite well to our moon, and in that sense they are every bit as unitary.

§ 33. Some writers go still further, and demand something not merely undivided but indivisible. G. Köpp[2] calls whatever is thought of as self-contained and incapable of dissection, whether we become acquainted with it through the senses or otherwise, an individual; and individuals which are to be numbered he calls ones, evidently using "one" here in the sense of "unit". BAUMANN too, when arguing for his view that external things do not present us with strict units on the ground that we are free to treat them as many, gives as a criterion of a strict unit that it must be incapable of dissection. Obviously, by tightening up its internal cohesion without limit, they hope to arrive at a criterion for their unit which is independent of any arbitrary way of regarding things. This attempt collapses because we are then left with practically nothing fit to be called a unit and to be numbered. The result is that we at once begin to retrace our steps, by giving as the criterion not that the thing itself should be incapable of dissection in fact, but that we should think of it as such. This brings us back once again to our way of regarding things, with all its fluctuations. And is it really of any advantage to think of things as being what they are not? On the contrary, any

---

[1] On the history of the word "unit" cf. Eucken, *Geschichte der philosophischen Terminologie*, pp. 122–23, 136, 220.

[2] *Schularithmetik*, pp. 5–6, Eisenach 1867.

conclusions drawn from a false assumption are liable to be false; while if there are no conclusions to be drawn from our unit's being incapable of dissection, why bother to assume that it is? If it does no harm, and in fact is actually necessary, to take our strict units none too strictly, what was the point of being strict? Unless perhaps all that was meant was that we should not think of the possibilities of dissection—as though lack of thought could get us anywhere! Besides, there are cases where we simply cannot avoid thinking of them, where a conclusion is actually based on the way in which a unit is made up of parts, as for instance in the problem: If there are 24 hours in one day, how many are there in three days?

Are units identical with one another?*

§ 34. Every attempt to define "one" as a property having thus failed, we must finally abandon the view that in designating a thing a unit we are adding to our description of it. We come back once again to our question: Why do we call things units, if "unit" is only another name for thing, if any and every thing is a unit or can be regarded as one? E. Schröder[1] gives as the reason, that the word is used for ascribing to the items that are to be numbered the necessary identity. But to begin with it is not easy to see why the words "thing" and "object" could not indicate this just as well. And further, it is natural to ask: Why do we ascribe identity to objects that are to be numbered? And is it only ascribed to them, or are they really identical? In any case, no two objects are *ever* completely identical. On the other hand, of course, we can practically always engineer some respect in which any two objects whatever agree. And with that we are back once more at our arbitrary way of regarding things, unless we are willing, regardless of truth, to ascribe to things an identity going beyond any they actually possess. In actual fact, many writers

[1] Op. cit., p. 5.

* [*gleich*. See footnote on p. IIe above.]

do call units identical without any qualification. Hobbes[1] states that: "Number in the absolute sense in mathematics presupposes units identical* one with another out of which it is formed." Hume[2] holds the component parts of quantity and number to be entirely similar. Thomae[3] calls the individual member of his set a unit, and says in so many words that units are identical with each other—though we might with as much or even more justice say that the individual members of his set must be different from each other. Now what has this alleged identity to do with number? The properties which serve to distinguish things from one another are, when we are considering their Number, immaterial and beside the point. That is why we want to keep them out of it. But we shall not succeed along the present lines. For suppose that we do, as Thomae demands, "abstract from the peculiarities of the individual members of a set of items", or "disregard, in considering separate things, those characteristics which serve to distinguish them". In that event we are not left, as Lipschitz maintains, with "the concept of the Number of the things considered"; what we get is rather a general concept under which the things in question fall. The things themselves do not in the process lose any of their special characteristics. If, for example, in considering a white cat and a black cat, I disregard the properties which serve to distinguish them, then I get presumably the concept "cat". Even if I proceed to bring them both under this concept and call them, I suppose, units, the white one still remains white just the same, and the black black. I may not think about their colours, or I may propose to make no inference from their difference in this respect, but for all that the cats do not become colourless and they remain different precisely as before. The concept "cat", no doubt, which we have arrived at by abstraction,

---

[1] Baumann, op. cit., Vol. I, p. 242 [op. cit., Dial. I, p. 16 = Molesworth edn., p. 18].

[2] Baumann, op. cit., Vol. II, p. 568 [*Enquiry concerning Human Understanding*, Sect. XII, part iii, § 131].

[3] Op. cit., p. 1.

* [*aequales*]

no longer includes the special characteristics of either, but of it, for just that reason, there is only one.

§ 35. We cannot succeed in making different things identical simply by dint of operations with concepts. But even if we did, we should then no longer have things in the plural, but only one thing; for, as DESCARTES[1] says, the number (or better, the plurality) in things arises from their diversity. And as E. SCHRÖDER[2] justly observes: "That things should be numbered is a reasonable demand only where the objects submitted appear clearly distinguishable from one another (for example, spatially and temporally separated) and isolated in contrast with one another." It does actually happen at times that too great similarity, for instance of the uprights in a railing, does make numbering difficult. W. S. JEVONS[3] makes this point with unusual force: "Number is but another name for *diversity*. Exact identity is unity, and with difference arises plurality." And again (p. 157)*: "It has often been said that units are units in respect of being perfectly similar to each other; but though they may be perfectly similar in some respects, they must be different in at least one point, otherwise they would be incapable of plurality. If three coins were so similar that they occupied the same place at the same time, they would not be three coins, but one."

§ 36. However, the view that units must be different comes up, as soon transpires, against fresh difficulties. JEVONS defines a unit as "any object of thought which can be discriminated from every other object treated as a unit in the same problem." But this is to define unit in terms of itself,

---

[1] Baumann, op. cit., Vol. I, p. 103 [*Principia*, Part I, § 60].

[2] Op. cit., p. 3.

[3] *The principles of science*, 3rd edn., p. 156 [1874 edn., p. 175].

* [1874 edn., p. 176]

and the qualifying clause "which can be discriminated from every other object" fails to describe it any more precisely, because it goes without saying: we call them other objects simply and solely because we can discriminate them from the first mentioned. JEVONS[1] goes on:·"Whenever I use the symbol 5 I really mean

$$1 + 1 + 1 + 1 + 1,$$

and it is perfectly understood that each of these units is distinct from each other. If requisite I might mark them thus

$$1' + 1'' + 1''' + 1'''' + 1'''''.$$"

Certainly it is requisite to mark them differently, if they are different: otherwise the utmost confusion must result. For if a difference simply in the position in which the 1 appears were to be made to mean of itself a difference in the unit, this convention would have to be laid down as a rule without any exception, or else we should never know whether $1 + 1$ was to be taken to mean 2 or 1. Accordingly, we should have to give up the equation $1 = 1$ and we should never, to our embarrassment, be able to mark the same thing twice. That obviously will not do. If, however, we adopt the alternative plan, of assigning different symbols to different things, it is hard to see why we still retain in our symbols a common element; why not write, instead of

$$1' + 1'' + 1''' + 1'''' + 1''''',$$

simply

$$a + b + c + d + e\,?$$

But now the identity of the units has been completely lost, and it helps not at all to point out that they are to some extent similar. So our one slips through our fingers; we are left with the objects in all their particularity. The symbols

$$1', 1'', 1'''$$

tell the tale of our embarrassment. We must have identity—

---

[1] Op. cit., p. 162 [1874 edn., p. 182].

hence the 1; but we must have difference—hence the strokes; only unfortunately, the latter undo the work of the former.

§ 37. In other writers we meet with the same difficulty. In LOCKE[1] we read: "By the repeating . . . of the idea of an unit, and joining it to another unit, we make thereof one collective idea, marked by the name two. And whosoever can do this and proceed on, still adding one more to the last collective idea which he had of any number, and give a name to it, may count." LEIBNIZ[2] defines number as 1 and 1 and 1 or as units. HESSE[3] writes: "Anyone who can form for himself an idea of the unit which in algebra is expressed by the symbol 1, . . . can go on to conceive a second unit as good as the first, and then further units of the same sort. The union of the second with the first into a single whole yields the number 2."

In these passages, the relation between the meanings of the words "unit" and "one" should be noticed. LEIBNIZ understands by *unitas* a concept under which this one and that one and the other one fall, or as he also puts it: "Abstractum ab uno est *Unitas*.". LOCKE and HESSE seem to use unit and one to mean the same. Indeed LEIBNIZ, in the last analysis, does so too; for when he calls each individual object falling under his concept of *unitas* a *unum*, this word is being used to signify not the individual object but the concept under which they all fall.

§ 38. However, if confusion is not to become worse confounded, it is advisable to observe a strict distinction

[1] Baumann, op. cit., Vol. I, pp. 410-11 [*Essay*, Bk. II., cap. xvi, § 5].
[2] Baumann, op. cit., Vol. II, p. 3 [Erdmann edn., p. 53].
[3] *Vier Species* [Leipzig 1872], p. 2.

between unit and one. When we speak of "the number one", we indicate by means of the definite article a definite and unique object of scientific study. There are not divers numbers one, but only one. In 1 we have a proper name, which as such does not admit of a plural any more than "Frederick the Great" or "the chemical element gold". It is no accident, nor is it a notational inexactitude, that we write 1 without any strokes to mark differences. JEVONS would rewrite the equation

$$3 - 2 = 1$$

in some such way as this:

$$(1' + 1'' + 1''') - (1'' + 1''') = 1'.$$

But what would be the remainder of

$$(1' + 1'' + 1''') - (1'''' + 1''''') ?$$

Certainly not 1'. It follows, therefore, that on his view there would be not only distinct ones but also distinct twos and so on; for 1'''' + 1''''' could not be substituted for 1'' + 1'''. This puts us in a position to see quite clearly that number is not an agglomeration of things. Arithmetic would come to a dead stop, if we tried to introduce in place of the number one, which is always the same, different distinct things, however similar the symbols for them; yet to make the symbols identical would be, of course, a mistake, and surely we cannot suppose that the mainspring of arithmetic is a piece of faulty notation. It is therefore impossible to regard 1 as a symbol for different distinct objects, for Iceland, Aldebaran, Solon, and so on. The absurdity can be best brought out by taking the case of an equation which has three roots, namely 2, 5 and 4. Suppose now with JEVONS we write for 3

$$1' + 1'' + 1''';$$

and let us take 1' and 1'' and 1''' to be units, that is, still following JEVONS, to be the objects currently under consideration. It follows that 1' would here mean 2, and 1'' 5, and 1''' 4. Then would it not be more intelligible, instead of 1' + 1'' + 1''', to write

$$2 + 5 + 4?$$

Only concept words can form a plural. If, therefore, we speak of "units", we must be using the word not as equivalent to the proper name "one", but as a concept word. If this term "unit" means "object to be numbered", then number cannot be defined as units. But if we understand by "unit" a concept which includes under it the number one and nothing else, a plural makes no sense, and it becomes impossible once more to define number, with LEIBNIZ, as units, or as 1 and 1 and 1; for if "and" is used as in "Bunsen and Kirchhof", then 1 and 1 and 1 is not 3 but 1, just as gold and gold and gold is never anything else but gold. The plus symbol in

$$1 + 1 + 1 = 3$$

must, therefore, be interpreted differently from the "and" which we use in symbolizing a collection or a "collective idea".

§ 39. We are faced, therefore, with the following difficulty:

If we try to produce the number by putting together different distinct objects, the result is an agglomeration in which the objects contained remain still in possession of precisely those properties which serve to distinguish them from one another; and that is not the number. But if we try to do it in the other way, by putting together identicals, the result runs perpetually together into one and we never reach a plurality.

If we use 1 to stand for each of the objects to be numbered, we make the mistake of assigning the same symbol to different things. But if we provide the 1 with differentiating strokes, it becomes unusable for arithmetic.

The word "unit" is admirably adapted to conceal this difficulty; and that is the real, though no doubt unconscious, reason why we prefer it to the words "object" and "thing". We start by calling the things to be numbered "units", without detracting from their diversity; then subsequently the concept of putting together (or collecting, or uniting, or annexing, or whatever we choose to call it) transforms itself into that of arithmetical addition, while the concept word "unit" changes unperceived into the proper name "one". And there we have our identity. If I annex to the letter *a* first an *n* and then a *d*, anyone can easily see that that is not the number 3. If, however, I bring the letters *a*, *n* and *d* under the concept "unit", and now, instead of "*a* and *n* and *d*", say "a unit and a unit and a further unit" or "1 and 1 and 1", we are quite prepared to believe that this does give us the number 3. The difficulty is so well hidden under the word "unit", that those who have any suspicion of its existence must surely be few at most.

Here, indeed, is an artful manipulation of language worthy of MILL's censure; for this is no outward manifestation of an inward process of thought, but only the illusion of one. Here we really do have the impression that words devoid of thought must possess some mysterious power, if what is different is to be made identical simply by being called a unit.

Attempts to overcome the difficulty.

§ 40. We consider next some detailed views which represent attempts to overcome this difficulty, although they have not always been produced with that end clearly and consciously in view.

The first suggestion is to call for assistance on a certain property of time and space, as follows. One point of space, considered by itself, is absolutely indistinguishable from another, and so is a straight line, or a plane, or one of a number of congruent bodies or areas or line-segments: they are distinguishable only when conjoined as elements in a single

total intuition. Here, therefore, we seem to get identity
combined with distinguishability. With the parts of time, the
same applies. This is presumably why HOBBES[1] holds it for
hardly conceivable that the identity of units should result
from anything but the division of the continuum. As THOMAE[2]
puts it: "If we consider a set of individuals or units in space
and number them one after the other, for which time is
necessary, then, abstract as we will, there remain always as
discriminating marks of the units their different positions in
space and in the order of succession in time."

The first doubt that strikes us about any such view is that
then nothing would be numerable except what is spatial and
temporal. LEIBNIZ[3] long ago rebutted the view of the schoolmen that number results from the mere division of the continuum and cannot be applied to immaterial things. BAUMANN[4]
dissociates number emphatically from time: he claims that the
concept of the unit is thinkable even apart from time. JEVONS[5]
writes: "Three coins are three coins, whether we count them
successively or regard them all simultaneously. In many cases
neither time nor space is the ground of difference, but pure
quality alone enters. We can discriminate, for instance, the
weight, inertia, and hardness of gold as three qualities, though
none of these is before or after the other, either in space or
time. Every means of discrimination may be a source of
plurality." I would add that, if the objects numbered do not
follow one after another in actual fact, but it is only that they
are numbered one after another, then time cannot be the ground
of discrimination between them. For, if we are to be able to
number them one after another, we must already be in posses-

1 Baumann, op. cit., Vol. I, p. 242 [loc. cit., p. 45e above].
2 Op. cit., p. 1.
3 Baumann, op. cit., Vol. II, p. 2 [loc. cit., p. 31e above].
4 Op. cit., Vol. II, p. 668.
5 Op. cit., p. 157 [1874 edn., p. 176].

sion of distinguishing marks. Time is only a psychological necessity for numbering, it has nothing to do with the concept of number. We do represent objects which are non-spatial and non-temporal by spatial or temporal points, and this may perhaps be of advantage in carrying out the procedure of numbering; but it presupposes, fundamentally, that the concept of number is applicable to the non-spatial and the non-temporal.

§ 41. But further, supposing we do disregard all distinguishing marks except those of space and time, do we then really succeed in combining distinguishability with identity? Not at all. We are not one step nearer a solution. Whether the objects are so much more similar or so much less is beside the point, if they have still to be kept separate in the end. I cannot here symbolize the individual points, or lines or whatever it may be, all alike by 1, any more than for purposes of geometry I can call them one and all A; in the one case as in the other, it is essential to distinguish between them. It is only considered in themselves, and neglecting their spatial relations, that points of space are identical with one another; if I am to think of them together, I am bound then to consider them in their collocation in space, or else they fuse irretrievably together into one. Now points taken together as a group may perhaps fall into some pattern or other like a constellation or may equally arrange themselves somehow or other on a straight line; and a group of identical segments may lie perhaps with their end-points adjacent so as to combine into a single segment or perhaps at a distance from one another. Patterns produced in this way can be completely different while the number of their elements remains the same. So that here once again we should have different distinct fives, sixes, and so forth. Points of time, again, are separated by time intervals, long or short, equal or unequal. All these are relationships which have absolutely nothing to do with number as such. Pervading them all is an admixture of some special element, which number in its general form leaves far behind. Even a single moment itself has something *sui generis*, which serves to distinguish it from, say, a point of space, and of which there is no trace in the concept of number.

§ 42. Another way out is to invoke instead of spatial or temporal order a more generalized concept of series, but this too fails of its object; for their positions in the series cannot be the basis on which we distinguish the objects, since they must already have been distinguished somehow or other, for us to have been able to arrange them in a series. Any such arrangement always presupposes relations between the objects, whether spatial or temporal or logical relations, or relations of pitch or what not, which serve to lead us on from one object to the next and which are necessarily bound up with distinguishing between them.

When HANKEL[1] speaks of our thinking or putting a thing once or twice or three times, this too seems to be an attempt to combine in the things to be numbered distinguishability with identity. But it is obvious too at once that it is not successful; for his ideas or intuitions of the same object must, if they are not to coalesce into one, be different in some way or other. Moreover we are, I imagine, entitled to speak of 45 million Germans without having first to have thought or put an average German 45 million times, which might be somewhat tedious.

§ 43. It is probably in order to avoid the difficulties which JEVONS runs into through making each symbol 1 mean one of the objects numbered, that E. SCHRÖDER allows it only to copy the object. The consequence is that he gives a definition not of number but only of numerals. To give his own words[2]: "To arrive at a symbol capable of expressing *how many*

---

[1] Op. cit., p. 1.
[2] Op. cit., pp. 5 ff.

of such units[1] are present, we direct our attention upon each one of them in turn *once*, and copy it by a stroke (a *one*), thus, 1; these ones we put in a row side by side, only linking them up to each other by the symbol + (plus) because otherwise 111, for example, would be read, following the usual number notation, as one hundred and eleven. In this way we get a symbol such as:

$$1 + 1 + 1 + 1 + 1,$$

the composition of which we can describe by saying:

'*A natural number is a sum of ones*'."

This passage shows that for SCHRÖDER number is a *symbol*. What the symbol expresses, which is what I have been calling number, is taken, with the words "how many of such units are present", as already known. Even by the word "one" he understands the symbol 1, not its meaning. The symbol + is introduced solely to serve as a visible mark, without any content of its own, for linking up the other symbols; only later does he define addition. He could indeed have put what he means more briefly by saying that we write down side by side as many symbols 1 as we have objects to be numbered, and link them up by the symbol +. Nought would be expressed by writing down nothing.

§ 44. To avoid carrying over into number the distinguishing marks of the things numbered, JEVONS[2] invokes abstraction: "There will now be little difficulty in forming a clear notion of the nature of numerical abstraction. It consists in abstracting the character of the difference from which plurality arises, retaining merely the fact. When I speak of

[1] Objects to be numbered.
[2] Op. cit., p. 158 [1874 edn., p. 177].

three men I need not at once specify the marks by which each may be known from each. Those marks must exist if they are really three men and not one and the same, and in speaking of them as many I imply the existence of the requisite differences. Abstract number, then, is *the empty form of difference*."

How are we to interpret this? Either we can abstract from the distinguishing properties of things before uniting them into a whole: or we can first form a whole and then abstract from the distinguishing properties. By the first method we should never get so far as to distinguish the things at all, and consequently could not retain the fact of the existence of the differences either; the second method seems to be what JEVONS intends. But by it we should never, it seems to me, arrive at a number like 10,000, for it is beyond our powers to grasp so many differences at once and retain the fact of their existence; while to go through them one after another is not enough, for the number would never be complete. We do our numbering in time, of course; but that does not give us the number itself, it only tells us the number of whatever it is we are numbering. Moreover, to tell us how to abstract is not, in any case, to give us a definition.

What are we to understand by "the empty form of difference"? Perhaps some proposition like

"*a* is different from *b*"

where *a* and *b* are left indefinite? Can this proposition be, say, the number 2? But does the proposition

"The Earth has two poles"

mean the same as

"The North Pole is different from the South"?

Obviously not. The second proposition could be true without the first being so, and vice versa. And for the number 1,000 we should then have as many as

$$\frac{1,000 \cdot 999}{1 \cdot 2}$$

such propositions, each stating a difference.

With the numbers 0 and 1 in particular, what Jevons says simply will not work. What is it, in fact, that we are supposed to abstract from, in order to get, for example, from the moon to the number 1? By abstraction we do indeed get certain concepts, viz. satellite of the Earth, satellite of a planet, non-self-luminous heavenly body, heavenly body, body, object. But in this series 1 is not to be met with; for it is no concept that the moon could fall under. In the case of 0, we have simply no object at all from which to start our process of abstracting. It is no good objecting that 0 and 1 are not numbers in the same sense as 2 and 3. What answers the question How many? is number, and if we ask, for example, "How many moons has this planet?", we are quite as much prepared for the answer 0 or 1 as for 2 or 3, and that without having to understand the question differently. No doubt there is something unique about 0, and about 1 too; but the same is true in principle of every whole number, only the bigger the number the less obvious it is. To make out of this a difference in kind is utterly arbitrary. What will not work with 0 and 1 cannot be essential to the concept of number.

Finally, by taking number to arise in this manner we do not by any means remove the difficulty encountered when we were considering the symbolization of 5 by

$$1' + 1'' + 1''' + 1'''' + 1'''''.$$

This notation agrees well with what Jevons says about the formation of number by abstraction; the strokes above the line, that is, indicate that a difference exists, without however specifying of what sort. But the mere existence of the difference is already enough, as we have seen, to produce on Jevons' view different distinct ones and twos and threes, which is utterly incompatible with the existence of arithmetic.

Solution of the difficulty.

§ 45. It is time now to survey what has been so far established and the questions which still remain unanswered.

Number is not abstracted from things in the way that colour, weight and hardness are, nor is it a property of things in the sense that they are. But when we make a statement of number*, what is that of which we assert something? This question remained unanswered.

Number is not anything physical, but nor is it anything subjective (an idea).

Number does not result from the annexing of thing to thing. It makes no difference even if we assign a fresh name after each act of annexation.

The terms "multitude", "set" and "plurality" are unsuitable, owing to their vagueness, for use in defining number.

In considering the terms one and unit, we left unanswered the question: How are we to curb the arbitrariness of our ways of regarding things, which threatens to obliterate every distinction between one and many?

Being isolated, being undivided, being incapable of dissection—none of these can serve as a criterion for what we express by the word "one".

If we call the things to be counted units, then the assertion that units are identical is, if made without qualification, false. That they are identical in this respect or that is true enough but of no interest. It is actually necessary that the things to be counted should be different if number is to get beyond 1.

We were thus forced, it seemed, to ascribe to units two contradictory qualities, namely identity and distinguishability.

A distinction must be drawn between one and unit. The word "one", as the proper name of an object of mathe-

* [See n. on p. 34⁰ supra.]

matical study, does not admit of a plural. Consequently, it is nonsense to make numbers result from the putting together of ones. The plus symbol in $1 + 1 = 2$ cannot mean such a putting together.

§ 46. It should throw some light on the matter to consider number in the context of a judgement which brings out its basic use. While looking at one and the same external phenomenon, I can say with equal truth both "It is a copse" and "It is five trees", or both "Here are four companies" and "Here are 500 men". Now what changes here from one judgement to the other is neither any individual object, nor the whole, the agglomeration of them, but rather my terminology. But that is itself only a sign that one concept has been substituted for another. This suggests as the answer to the first of the questions left open in our last paragraph, that the content of a statement of number is an assertion about a concept. This is perhaps clearest with the number o. If I say "Venus has o moons", there simply does not exist any moon or agglomeration of moons for anything to be asserted of; but what happens is that a property is assigned to the *concept* "moon of Venus", namely that of including nothing under it. If I say "the King's carriage is drawn by four horses", then I assign the number four to the concept "horse that draws the King's carriage".

It may be objected that a concept like "inhabitant of Germany" would then possess, in spite of there being no change in its defining characteristics, a property which varied from year to year, if statements of the number of inhabitants did really assert a property of it. In reply to this, it is enough to point out that objects too can change their properties without that preventing us from recognizing them as the same. In this case, however, we can actually give the explanation more precisely. The fact is that the concept "inhabitant of Germany" contains a time-reference as a variable element in it, or, to put it mathematically, is a function of the time.

Instead of "*a* is an inhabitant of Germany" we can say "*a* inhabits Germany", and this refers to the current date at the time. Thus in the concept itself there is already something fluid. On the other hand, the number belonging to the concept "inhabitant of Germany at New Year 1883, Berlin time" is the same for all eternity.

§ 47. That a statement of number should express something factual independent of our way of regarding things can surprise only those who think a concept is something subjective like an idea. But this is a mistaken view. If, for example, we bring the concept of body under that of what has weight, or the concept of whale under that of mammal, we are asserting something objective; but if the concepts themselves were subjective, then the subordination of one to the other, being a relation between them, would be subjective too, just as a relation between ideas is. It is true that at first sight the proposition

"All whales are mammals"

seems to be not about concepts but about animals; but if we ask which animal then are we speaking of, we are unable to point to any one in particular. Even supposing a whale is before us, our proposition still does not state anything about it. We cannot infer from it that the animal before us is a mammal without the additional premiss that it is a whale, as to which our proposition says nothing. As a general principle, it is impossible to speak of an object without in some way designating or naming it; but the word "whale" is not the name of any individual creature. If it be replied that what we are speaking of is not, indeed, an individual definite object, but nevertheless an indefinite object, I suspect that "indefinite object" is only another term for concept, and a poor one at that, being self-contradictory. However true it may be that our proposition can only be verified by observing particular animals, that proves nothing as to its content; to decide what it is about, we do not need to know whether it is true or not, nor for what reasons we believe it to be true. If, then, a concept is something objective, an assertion about a concept can have for its part a factual content.

§ 48. Several examples given earlier gave the false impression that different numbers may belong to the same thing. This is to be explained by the fact that we were there taking objects to be what has number. As soon as we restore possession to the rightful owner, the concept, numbers reveal themselves as no less mutually exclusive in their own sphere than colours are in theirs.

We now see also why there is a temptation to suggest that we get the number by abstraction from the things. What we do actually get by such means is the concept, and in this we then discover the number. Thus abstraction does genuinely often precede the formation of a judgement of number. It would be an analogous confusion to maintain that the way to acquire the concept of fire risk is to build a frame house, with timber gables, thatched roof and leaky chimneys.

The concept has a power of collecting together far superior to the unifying power of synthetic apperception. By means of the latter it would not be possible to join the inhabitants of Germany together into a whole; but we can certainly bring them all under the concept "inhabitant of Germany" and number them.

The wide range of applicability of number also now becomes explicable. Not without reason do we feel it puzzling that we should be able to assert the same predicate of physical and mental phenomena alike, of the spatial and temporal and of the non-spatial and non-temporal. But then, this simply is not what occurs with statements of number any more than elsewhere; numbers are assigned only to the concepts, under which are brought both the physical and mental alike, both the spatial and temporal and the non-spatial and non-temporal.

§ 49. Corroboration for our view is to be found in SPINOZA,[1] where he writes: "I answer that a thing is called one or single simply with respect to its existence, and not with respect to its essence; for we only think of things in terms of number after they have first been reduced to a common genus. For example, a man who holds in his hand a sesterce and a dollar will not think of the number two unless he can cover his sesterce and his dollar with one and the same name, viz., piece of silver, or coin; then he can affirm that he has two pieces of silver, or two coins; since he designates by the name piece of silver or coin not only the sesterce but also the dollar." Unfortunately, he goes on: "From this it is clear, therefore, that nothing is called one or single except when some other thing has first been conceived which, as has been said, matches it", and he holds further that we cannot properly call God one or single, because it would be impossible for us to form an abstract concept of his essence. Here he makes the mistake of supposing that a concept can only be acquired by direct abstraction from a number of objects. We can, on the contrary, arrive at a concept equally well by starting from defining characteristics; and in such a case it is possible for nothing to fall under it. If this did not happen, we should never be able to deny existence, and so the assertion of existence too would lose all content.

§ 50. E. SCHRÖDER[2] calls attention to the fact that, if we are to be able to speak of the frequency of a thing, the name of the thing concerned must always be a *generic name*, a general concept word or *notio communis*; "So soon, that is, as we picture an object complete—with all its properties and in

---

[1] Baumann, op. cit., Vol. I, p. 169 [*Epistolae doctorum quorundam virorum*, No. 50 *ad* J. Jelles].

[2] Op. cit., p. 6.

all its relations, it will present itself as unique in the universe, and there will no longer be anything to match it. The name of the object takes on at once the character of a *proper name* (*nomen proprium*), and the object itself cannot be thought of as one which is found more than once. But observe that this holds good not only of *concrete* objects, but generally of anything and everything, even where the idea of it arises through *abstractions*, provided only that this idea contains in it sufficient elements to constitute the thing concerned a *completely* determinate thing. . .". For a thing to be numbered "first becomes possible in so far as, for that purpose, we disregard or *abstract from* some of its peculiar characteristics and relations, which distinguish it from all other things; this has the effect of turning what was the name of the thing into a concept applicable to more than one thing."

§ 51. What is true in this account is wrapped up in such distorted and misleading language, that we are obliged to straighten it out and sort the wheat from the chaff. To start with, it will not do to call a general concept word the name of a thing. That leads straight to the illusion that the number is a property of a thing. The business of a general concept word is precisely to signify a concept. Only when conjoined with the definite article or a demonstrative pronoun can it be counted as the proper name of a thing, but in that case it ceases to count as a concept word. The name of a thing is a proper name. An object, again, is not found more than once, but rather, more than one object falls under the same concept. That a concept need not be acquired by abstraction from the things which fall under it has already been pointed out in criticizing Spinoza. Here I will add that a concept does not cease to be a concept simply because only one single thing falls under it, which thing, accordingly, is completely determined by it. It is to concepts of just this kind (for example, satellite of the Earth) that the number 1 belongs, which is a number in the same sense as 2 and 3. With a concept the question is always whether anything, and if so what, falls under it. With a proper name such questions make no sense. We should not be deceived by the fact that language makes use of proper names, for instance Moon, as concept words, and vice versa; this does not affect the distinction between the two. As soon as a word is used with the indefinite article or in the plural without any article, it is a concept word.

§ 52. Further confirmation of the view that number is assigned to concepts is to be found in idiom; just as in English we can speak of "three barrel", so in German we speak generally of "ten man", "four mark" and so on. The use of the singular here may indicate that the concept is intended, not the thing. The advantage of this way of speaking is particularly noticeable in the case of the number 0. Elsewhere, it must be admitted, our ordinary language does assign number not to concepts but to objects: we speak of "the number of the bales" just as we do of "the weight of the bales". Thus on the face of it we are talking about objects, whereas really we are intending to assert something of a concept. This usage is confusing. The construction in "four thoroughbred horses" fosters the illusion that "four" modifies the concept "thoroughbred horse" in just the same way as "thoroughbred" modifies the concept "horse." Whereas in fact only "thoroughbred" is a characteristic used in this way; the word "four" is used to assert something of a concept.

§ 53. By properties which are asserted of a concept I naturally do not mean the characteristics which make up the concept. These latter are properties of the things which fall under the concept, not of the concept. Thus "rectangular" is not a property of the concept "rectangular triangle"; but the proposition that there exists no rectangular equilateral rectilinear triangle does state a property of the concept "rectangular equilateral rectilinear triangle"; it assigns to it the number nought.

In this respect existence is analogous to number. Affirmation of existence is in fact nothing but denial of the number nought. Because existence is a property of concepts the ontological argument for the existence of God breaks down. But oneness* is not a component characteristic of the concept "God" any more than existence is. Oneness cannot be used in the definition of this concept any more than the solidity of a house, or its commodiousness or desirability, can be used in building it along with the beams, bricks and mortar. However, it would be wrong to conclude that it is in principle impossible ever to deduce from a concept, that is, from its component characteristics, anything which is a property of the concept. Under certain conditions this is possible, just as we can occasionally infer the durability of a building from the type of stone used in building it. It would therefore be going too far to assert that we can never infer from the component characteristics of a concept to oneness or to existence; what is true is, that this can never be so direct a matter as it is to assign some component of a concept as a property to an object falling under it.

It would also be wrong to deny that existence and oneness can ever themselves be component characteristics of a concept. What is true is only that they are not components of those particular concepts to which language might tempt us to ascribe them. If, for example, we collect under a single concept all concepts under which there falls only one object, then oneness is a component characteristic of this new concept. Under it would fall, for example, the concept "moon of the Earth", though not the actual heavenly body called by this name. In this way we can make one concept fall under another higher or, so to say, second order concept. This relationship, however, should not be confused with the subordination of species to genus.

§ 54. It now becomes possible to give a satisfactory definition of the term "unit". E. SCHRÖDER writes, on p. 7 of his text book already referred to: "This generic name or

* [I.e. the character of being single or unique, called by theologians "unity".]

concept will be called the denomination of the number formed by the method given, and constitutes, in effect, what is meant by its unit."

Why not, in fact, adopt this very apt suggestion, and call a concept the unit relative to the Number which belongs to it? We can then achieve a sense for the assertions made about the unit, that it is isolated from its environment and is indivisible. For it is the case that the concept, to which the number is assigned, does in general isolate in a definite manner what falls under it. The concept "letters in the word three" isolates the *t* from the *h*, the *h* from the *r*, and so on. The concept "syllables in the word three" picks out the word as a whole, and as indivisible in the sense that no part of it falls any longer under that same concept. Not all concepts possess this quality. We can, for example, divide up something falling under the concept "red" into parts in a variety of ways, without the parts thereby ceasing to fall under the same concept "red". To a concept of this kind no finite number will belong. The proposition asserting that units are isolated and indivisible can, accordingly, be formulated as follows:

Only a concept which isolates what falls under it in a definite manner, and which does not permit any arbitrary division of it into parts, can be a unit relative to a finite Number.

It will be noticed, however, that "indivisibility" here has a special meaning.

We can now easily solve the problem of reconciling the identity of units with their distinguishability. The word "unit" is being used here in a double sense. The units are identical if the word has the meaning just explained. In the proposition "Jupiter has four moons", the unit is "moon of Jupiter". Under this concept falls moon I, and likewise also moon II, and moon III too, and finally moon IV. Thus we can say: the unit to which I relates is identical with the unit to which II relates, and so on. This gives us our identity.

But when we assert the distinguishability of units, we mean that the things numbered are distinguishable.

IV. The concept of Number.

Every individual number is a self-subsistent object.

§ 55. Now that we have learned that the content of a statement of number is an assertion about a concept, we can try to complete the Leibnizian definitions of the individual numbers by giving the definitions of o and of 1.

It is tempting to define o by saying that the number o belongs to a concept if no object falls under it. But this seems to amount to replacing o by "no", which means the same. The following formulation is therefore preferable: the number o belongs to a concept, if the proposition that *a* does not fall under that concept is true universally, whatever *a* may be.

Similarly we could say: the number 1 belongs to a concept *F*, if the proposition that *a* does not fall under *F* is not true universally, whatever *a* may be, and if from the propositions

"*a* falls under *F*" and "*b* falls under *F*"

it follows universally that *a* and *b* are the same.

It remains still to give a general definition of the step from any given number to the next. Let us try the following formulation: the number $(n + 1)$ belongs to a concept *F*, if there is an object *a* falling under *F* and such that the number *n* belongs to the concept "falling under *F*, but not *a*".

§ 56. These definitions suggest themselves so spontaneously in the light of our previous results, that we shall have to go into the reasons why they cannot be reckoned satisfactory.

The most likely to cause misgivings is the last; for strictly speaking we do not know the sense of the expression "the number *n* belongs to the concept *G*" any more than we do that of the expression "the number (*n* + 1) belongs to the concept *F*". We can, of course, by using the last two definitions together, say what is meant by

"the number 1 + 1 belongs to the concept *F*"

and then, using this, give the sense of the expression

"the number 1 + 1 + 1 belongs to the concept *F*"

and so on; but we can never—to take a crude example—decide by means of our definitions whether any concept has the number Julius Caesar belonging to it, or whether that same familiar conqueror of Gaul is a number or is not. Moreover we cannot by the aid of our suggested definitions prove that, if the number *a* belongs to the concept *F* and the number *b* belongs ·to the same concept, then necessarily *a* = *b*. Thus we should be unable to justify the expression "*the* number which belongs to the concept *F*", and therefore should find it impossible in general to prove a numerical identity, since we should be quite unable to achieve a determinate number. It is only an illusion that we have defined 0 and 1; in reality we have only fixed the sense of the phrases

"the number 0 belongs to"

"the number 1 belongs to";

but we have no authority to pick out the 0 and 1 here as self-subsistent objects that can be recognized as the same again.

§ 57. It is time to get a clearer view of what we mean by our expression "the content of a statement of number is an assertion about a concept". In the proposition "the number 0 belongs to the concept *F*", 0 is only an element in the predicate (taking the concept *F* to be the real subject). For this reason I have avoided calling a number such as 0 or 1 or 2 a *property* of a concept. Precisely because it forms only an element in what is asserted, the individual number shows itself for what it is, a self-subsistent object. I have already drawn attention above to the fact that we speak of "the number 1", where the definite article serves to class it as an object. In arithmetic this self-subsistence comes out at every turn, as for example in the identity $1 + 1 = 2$. Now our concern here is to arrive at a concept of number usable for the purposes of science; we should not, therefore, be deterred by the fact that in the language of everyday life number appears also in attributive constructions. That can always be got round. For example, the proposition "Jupiter has four moons" can be converted into "the number of Jupiter's moons is four". Here the word "is" should not be taken as a mere copula, as in the proposition "the sky is blue". This is shown by the fact that we can say: "the number of Jupiter's moons is the number four, or 4" Here "is" has the sense of "is identical with" or "is the same as". So that what we have is an identity, stating that the expression "the number of Jupiter's moons" signifies the same object as the word "four". And identities are, of all forms of proposition, the most typical of arithmetic. It is no objection to this account that the word "four" contains nothing about Jupiter or moons. No more is there in the name "Columbus" anything about discovery or about America, yet for all that it is the same man that we call Columbus and the discoverer of America.

§ 58. A possible criticism is, that we are not able to form of this object which we are calling Four or the Number of Jupiter's moons any sort of idea[1] at all which would make it something self-subsistent. But that is not the fault of the self-subsistence we have ascribed to the number. It is easy, I know, to suppose that in our idea of four dots on a die there is to be found something which corresponds to the word "four"; but that is a misapprehension. We have only to think of a green field, and try whether the idea alters when we replace the indefinite article by the number word "one"; nothing fresh is added—whereas with the word "green", there really is in the idea something which corresponds to it. If we

---

[1] "Idea" in the sense of something like a picture.

imagine the printed word "gold", we shall not immediately think of any number in connexion with it. If we now ask ourselves how many letters it contains, the number 4 is the result; yet the idea does not become in consequence any more definite, but may remain completely unaltered. Where we discover the number is precisely in the freshly added concept "letter in the word gold". In the case of the four dots on the die, the matter is rather more obscured, because the concept thrusts itself upon us so immediately, owing to the similarity of the dots, that we scarcely notice its intervention. We can form no idea of the number either as a self-subsistent object or as a property in an external thing, because it is not in fact either anything sensible or a property of an external thing. But the point is clearest in the case of the number 0; we shall try in vain to form an idea of 0 visible stars. We can, of course, think of a sky entirely overcast with clouds; but in this there is nothing to correspond to the word "star" or to 0. All we succeed in imagining is a situation where the natural judgement to make would be: No star is now to be seen.

§ 59. It may be that every word calls up some sort of idea in us, even a word like "only"; but this idea need not correspond to the content of the word; it may be quite different in different men. The sort of thing we do is to imagine a situation where some proposition in which the word occurs would be called for; or it may happen that the spoken word recalls the written word to our memory.

Nor does this happen only in the case of particles. There is not the slightest doubt that we can form no idea of our distance from the sun. For even although we know the rule that we must multiply a measuring rod so and so many times, we still fail in every attempt to construct by its means a picture approximating even faintly to what we want. Yet this is no reason for doubting the correctness of the calculation which established the distance, not does it prevent us in any way from taking that distance as a fact upon which to base further inferences.

§ 60. Even so concrete a thing as the Earth we are unable to imagine as we know it to be; instead, we content ourselves with a ball of moderate size, which serves us as a symbol for the Earth, though we know quite well it is very different from it. Thus even although our idea often fails entirely to coincide with what we want, we still make judgements about an object such as the Earth with considerable certainty, even where its size is in point.

Time and time again we are led by our thought beyond the scope of our imagination, without thereby forfeiting the support we need for our inferences. Even if, as seems to be the case, it is impossible for men such as we are to think without ideas, it is still possible for their connexion with what we are thinking of to be entirely superficial, arbitrary and conventional.

That we can form no idea of its content is therefore no reason for denying all meaning to a word, or for excluding it from our vocabulary. We are indeed only imposed on by the opposite view because we will, when asking for the meaning of a word, consider it in isolation, which leads us to accept an idea as the meaning. Accordingly, any word for which we can find no corresponding mental picture appears to have no content. But we ought always to keep before our eyes a complete proposition. Only in a proposition have the words really a meaning. It may be that mental pictures float before us all the while, but these need not correspond to the logical elements in the judgement. It is enough if the proposition taken as a whole has a sense; it is this that confers on its parts also their content.

This observation is destined, I believe, to throw light

on quite a number of difficult concepts, among them that of the infinitesimal,[1] and its scope is not restricted to mathematics either.

The self-subsistence which I am claiming for number is not to be taken to mean that a number word signifies something when removed from the context of a proposition, but only to preclude the use of such words as predicates or attributes, which appreciably alters their meaning.

§ 61. But, it will perhaps be objected, even if the Earth is really not imaginable, it is at any rate an external thing, occupying a definite place; but where is the number 4? It is neither outside us nor within us. And, taking those words in their spatial sense, that is quite correct. To give spatial co-ordinates for the number 4 makes no sense; but the only conclusion to be drawn from that is, that 4 is not a spatial object, not that it is not an object at all. Not every object has a place. Even our ideas[2] are in this sense not within us (beneath our skin); beneath the skin are nerve-ganglia, blood corpuscles and things of that sort, but not ideas. Spatial predicates are not applicable to them: an idea is neither to the right nor to the left of another idea; we cannot give distances between ideas in millimetres. If we still say they are within us, then we intend by this to signify that they are subjective.

Yet even granted that what is subjective has no position in space, how is it possible for the number 4, which is objective, not to be anywhere? Now I contend that there is no contradiction in this whatever. It is a fact that the number 4 is exactly the same for everyone who deals with it; but that has nothing to do with being spatial. Not every objective object* has a place.

---

[1] The problem here is not, as might be thought, to produce a segment bounded by two distinct points whose length is dx, but rather to define the sense of an identity of the type

$$df(x) = g(x)dx$$

[2] Understanding this word in its purely psychological, not in its psycho-physical, sense.

* [*objektiver Gegenstand*]

To obtain the concept of Number, we must fix the sense of a numerical identity.

§ 62. How, then, are numbers to be given to us, if we cannot have any ideas or intuitions of them? Since it is only in the context of a proposition that words have any meaning, our problem becomes this: To define the sense of a proposition in which a number word occurs. That, obviously, leaves us still a very wide choice. But we have already settled that number words are to be understood as standing for self-subsistent objects. And that is enough to give us a class of propositions which must have a sense, namely those which express our recognition of a number as the same again. If we are to use the symbol *a* to signify an object, we must have a criterion for deciding in all cases whether *b* is the same as *a*, even if it is not always in our power to apply this criterion. In our present case, we have to define the sense of the proposition

"the number which belongs to the concept *F* is the same
as that which belongs to the concept *G*";

that is to say, we must reproduce the content of this proposition in other terms, avoiding the use of the expression

"the Number which belongs to the concept *F*".

In doing this, we shall be giving a general criterion for the identity of numbers. When we have thus acquired a means of arriving at a determinate number and of recognizing it again as the same, we can assign it a number word as its proper name.

§ 63. HUME[1] long ago mentioned such a means: "When two numbers are so combined as that the one has always an unit answering to every unit of the other, we pronounce them equal." This opinion, that numerical equality or identity

---

[1] Baumann, op. cit., Vol. II, p. 565 *Treatise*, Bk. I, Part iii, Sect. 1].

must be defined in terms of one-one correlation, seems in recent years to have gained widespread acceptance among mathematicians.[1] But it raises at once certain logical doubts and difficulties, which ought not to be passed over without examination.

It is not only among numbers that the relationship of identity is found. From which it seems to follow that we ought not to define it specially for the case of numbers. We should expect the concept of identity to have been fixed first, and that then, from it together with the concept of Number, it must be possible to deduce when Numbers are identical with one another, without there being need for this purpose of a special definition of numerical identity as well.

As against this, it must be noted that for us the concept of Number has not yet been fixed, but is only due to be determined in the light of our definition of numerical identity. Our aim is to construct the content of a judgement which can be taken as an identity such that each side of it is a number. We are therefore proposing not to define identity specially for this case, but to use the concept of identity, taken as already known, as a means for arriving at that which is to be regarded as being identical. Admittedly, this seems to be a very odd kind of definition, to which logicians have not yet paid enough attention; but that it is not altogether unheard of, may be shown by a few examples.

§ 64. The judgement "line *a* is parallel to line *b*", or, using symbols,

$$a \mathbin{/\mkern-3mu/} b,$$

can be taken as an identity. If we do this, we obtain the concept of direction, and say: "the direction of line *a* is identical with the direction of line *b*". Thus we replace the symbol // by

---

[1] Cf. E. Schröder, op. cit., pp. 7–8; E. Kossak, *Die Elemente der Arithmetik, Programm des Friedrichs-Werder'schen Gymnasiums*, Berlin 1872, p. 16; G. Cantor, *Grundlagen einer allgemeinen Mannichfaltigkeitslehre*, Leipzig 1883.

the more generic symbol $=$, through removing what is specific in the content of the former and dividing it between a and b. We carve up the content in a way different from the original way, and this yields us a new concept. Often, of course, we conceive of the matter the other way round, and many authorities define parallel lines as lines whose directions are identical. The proposition that "straight lines parallel to the same straight line are parallel to one another" can than be very conveniently proved by invoking the analogous proposition about things identical with the same thing. Only the trouble is, that this is to reverse the true order of things. For surely everything geometrical must be given originally in intuition. But now I ask whether anyone has an intuition of the direction of a straight line. Of a straight line, certainly; but do we distinguish in our intuition between this straight line and something else, its direction? That is hardly plausible. The concept of direction is only discovered at all as a result of a process of intellectual activity which takes its start from the intuition. On the other hand, we do have an idea of parallel straight lines. Our convenient proof is only made possible by surreptitiously assuming, in our use of the word "direction", what was to be proved; for if it were false that "straight lines parallel to the same straight line are parallel to one another", then we could not transform $a \,/\,/\, b$ into an identity.

We can obtain in a similar way from the parallelism of planes a concept corresponding to that of direction in the case of straight lines; I have seen the name "orientation"* used for this. From geometrical similarity is derived the concept of shape, so that instead of "the two triangles are similar" we say "the two triangles are of identical shape" or "the shape of the one is identical with that of the other". It is possible to derive yet another concept in this way, to which no name has yet been given, from the collineation of geometrical forms.

* [*Stellung*]

§ 65. Now in order to get, for example, from parallelism[1] to the concept of direction, let us try the following definition:
The proposition

"line *a* is parallel to line *b*"

is to mean the same as

"the direction of line *a* is identical with the direction of line *b*".

This definition departs to some extent from normal practice, in that it serves ostensibly to adapt the relation of identity, taken as already known, to a special case, whereas in reality it is designed to introduce the expression "the direction of line *a*", which only comes into it incidentally. It is this that gives rise to a second doubt—are we not liable, through using such methods, to become involved in conflict with the well-known laws of identity? Let us see what these are. As analytic truths they should be capable of being derived from the concept itself alone. Now LEIBNIZ's[2] definition is as follows:

"Things are the same as each other, of which one can be substituted for the other without loss of truth".*

This I propose to adopt as my own definition of identity. Whether we use "the same", as LEIBNIZ does, or "identical", is not of any importance. "The same" may indeed be thought to refer to complete agreement in all respects, "identical"** only to agreement in this respect or that; but we can adopt a form of expression such that this distinction vanishes. For example, instead of "the segments are identical in length", we can say "the length of the segments is identical" or "the same", and instead of "the surfaces are identical in colour", "the colour of the surfaces is identical". And this is the way in which the word has been used in the examples above.

---

[1] I have chosen to discuss here the case of parallelism, because I can express myself less clumsily and make myself more easily understood. The argument can readily be transferred in essentials to apply to the case of numerical identity.

[2] *Non inelegans specimen demonstrandi in abstractis* (Erdmann edn., p. 94).

* [*Eadem sunt, quorum unum potest substitui alteri salva veritate.*]

** [Still more "equal" or "similar", which the German *gleich* can also mean.]

Now, it is actually the case that in universal substitutability all the laws of identity are contained.

In order, therefore, to justify our proposed definition of the direction of a line, we should have to show that it is possible, if line *a* is parallel to line *b*, to substitute
                    "the direction of *b*"
everywhere for
                    "the direction of *a*".
This task is made simpler by the fact that we are being taken initially to know of nothing that can be asserted about the direction of a line except the one thing, that it coincides with the direction of some other line. We should thus have to show only that substitution was possible in an identity of this one type, or in judgement-contents containing such identities as constituent elements.[1] The meaning of any other type of assertion about directions would have first of all to be defined, and in defining it we can make it a rule always to see that it must remain possible to substitute for the direction of any line the direction of any line parallel to it.

§ 66. But there is still a third doubt which may make us suspicious of our proposed definition. In the proposition

"the direction of *a* is identical with the direction of *b*"

the direction of *a* plays the part of an object,[2] and our definition affords us a means of recognizing this object as the same again, in case it should happen to crop up in some other guise, say as the direction of *b*. But this means does not provide for all

---

[1] In a hypothetical judgement, for example, an identity of directions might occur as antecedent or consequent.

[2] This is shown by the definite article. A concept is for me that which can be predicate of a singular judgement-content, an object that which can be subject of the same. If in the proposition
          "the direction of the axis of the telescope is identical with the direction
                    of the Earth's axis"
we take the direction of the axis of the telescope as subject, then the predicate is "identical with the direction of the Earth's axis". This is a concept. But the direction of the Earth's axis is only an element in the predicate; it, since it can also be made the subject, is an object.

cases. It will not, for instance, decide for us whether England is the same as the direction of the Earth's axis—if I may be forgiven an example which looks nonsensical. Naturally no one is going to confuse England with the direction of the Earth's axis; but that is no thanks to our definition of direction. That says nothing as to whether the proposition

"the direction of a is identical with q"

should be affirmed or denied, except for the one case where q is given in the form of "the direction of b". What we lack is the concept of direction; for if we had that, then we could lay it down that, if q is not a direction, our proposition is to be denied, while if it is a direction, our original definition will decide whether it is to be denied or affirmed. So the temptation is to give as our definition:

q is a direction, if there is a line b whose direction is q.

But then we have obviously come round in a circle. For in order to make use of this definition, we should have to know already in every case whether the proposition

"q is identical with the direction of b"

was to be affirmed or denied.

§ 67. If we were to try saying: q is a direction if it is introduced by means of the definition set out above, then we should be treating the way in which the object q is introduced as a property of q, which it is not. The definition of an object does not, as such, really assert anything about the object, but only lays down the meaning of a symbol. After this has been done, the definition transforms itself into a judgement, which does assert about the object; but now it no longer introduces the object, it is exactly on a level with other assertions made about it. If, moreover, we were to adopt this way out, we should have to be presupposing that an object can only be given in one single way; for otherwise it would not follow, from the fact that q *was* not introduced by means of our definition, that it *could* not have been introduced by means of it. All identities would then amount simply to this, that whatever is given to us in the same way is to be reckoned as the same. This, however, is a principle so obvious and so sterile as not to be worth stating. We could not, in fact, draw from it any conclusion which was not the same as one of our premisses. Why is it, after all, that we are able to make use of identities with such significant results in such divers fields? Surely it is rather because we are able to recognize something as the same again even although it is given in a different way.

§ 68. Seeing that we cannot by these methods obtain any concept of direction with sharp limits to its application, nor therefore, for the same reasons, any satisfactory concept of Number either, let us try another way. If line *a* is parallel to line *b*, then the extension of the concept "line parallel to line *a*" is identical with the extension of the concept "line parallel to line *b*"; and conversely, if the extensions of the two concepts just named are identical, then *a* is parallel to *b*. Let us try, therefore, the following type of definition:

the direction of line *a* is the extension of the concept "parallel to line *a*";
the shape of triangle *t* is the extension of the concept "similar to triangle *t*".

To apply this to our own case of Number, we must substitute for lines or triangles concepts, and for parallelism or similarity the possibility of correlating one to one the objects which fall under the one concept with those which fall under the other. For brevity, I shall, when this condition is satisfied, speak of the concept F being *equal** to the concept G; but I must ask that this word be treated as an arbitrarily selected symbol, whose meaning is to be gathered, not from its etymology, but from what is here laid down.

My definition is therefore as follows:

the Number which belongs to the concept F is the

* [*Gleichzahlig*—an invented word, literally "identinumerate" or "taut-arithmic"; but these are too clumsy for constant use. Other translators have used "equinumerous"; "equinumerate" would be better. Later writers have used "similar" in this connexion (but as a predicate of "class" not of "concept").]

extension[1] of the concept "equal to the concept
F".

§ 69. That this definition is correct will perhaps be
hardly evident at first. For do we not think of the extensions
of concepts as something quite different from numbers? How
we do think of them emerges clearly from the basic assertions
we make about them. These are as follows:

    1. that they are identical,
    2. that one is wider than the other.

But now the proposition:

the extension of the concept "equal to the concept *F*"
is identical with the extension of the concept "equal
to the concept *G*"

is true if and only if the proposition

"the same number belongs to the concept *F* as to the
concept *G*"

is also true. So that here there is complete agreement.

Certainly we do not say that one number is wider than
another, in the sense in which the extension of one concept
is wider than that of another; but then it is also quite impossible for a case to occur where

the extension of the concept "equal to the concept *F*"

would be wider than

---

[1] I believe that for."extension of the concept" we could write simply "concept". But this would be open to the two objections:

    1. that this contradicts my earlier statement that the individual numbers
are objects, as is indicated by the use of the definite article in expressions
like "the number two" and by the impossibility of speaking of ones, twos, etc.
in the plural, as also by the fact that the number constitutes only an element
in the predicate of a statement of number;

    2. that concepts can have identical extensions without themselves co-
inciding.

I am, as it happens, convinced that both these objections can be met; but to do
this would take us too far afield for present purposes. I assume that it is known
what the extension of a concept is.

the extension of the concept "equal to the concept G". For on the contrary, when all concepts equal to G are also equal to F, then conversely also all concepts equal to F are equal to G. "Wider" as used here must not, of course, be confused with "greater" as used of numbers.

Another type of case is, I admit, conceivable, where the extension of the concept "equal to the concept F" might be wider or less wide than the extension of some other concept, which then could not, on our definition, be a Number; and it is not usual to speak of a Number as wider or less wide than the extension of a concept; but neither is there anything to prevent us speaking in this way, if such a case should ever occur.

Our definition completed and its worth proved.

§ 70. Definitions show their worth by proving fruitful. Those that could just as well be omitted and leave no link missing in the chain of our proofs should be rejected as completely worthless.

Let us try, therefore, whether we can derive from our definition of the Number which belongs to the concept F any of the well-known properties of numbers. We shall confine ourselves here to the simplest.

For this it is necessary to give a rather more precise account still of the term "equality". "Equal" we defined in terms of one-one correlation, and what must now be laid down is how this latter expression is to be understood, since it might easily be supposed that it had something to do with intuition.

We will consider the following example. If a waiter wishes to be certain of laying exactly as many knives on a table as plates, he has no need to count either of them; all he has to do is to lay immediately to the right of every plate a knife, taking care that every knife on the table lies immediately to the right of a plate. Plates and knives are thus correlated one to one, and that by the identical spatial relationship. Now if in the proposition

"a lies immediately to the right of A"

we conceive first one and then another object inserted in place of a and again of A, then that part of the content which remains unaltered throughout this process constitutes the essence of the relation. What we need is a generalization of this.

If from a judgement-content which deals with an object a and an object b we subtract a and b, we obtain as remainder a relation-concept which is, accordingly, incomplete at two points. If from the proposition

"the Earth is more massive than the Moon"

we subtract "the Earth", we obtain the concept "more massive than the Moon". If, alternatively, we subtract the object, "the Moon", we get the concept "less massive than the Earth". But if we subtract them both at once, then we are left with a relation-concept, which taken by itself has no [assertible] sense any more than a simple concept has: it has always to be completed in order to make up a judgement-content. It can however be completed in different ways: instead of Earth and Moon I can put, for example, Sun and Earth, and this *eo ipso* effects the subtraction.

Each individual pair of correlated objects stands to the relation-concept much as an individual object stands to the concept under which it falls—we might call them the subject of the relation-concept. Only here the subject is a composite one. Occasionally, where the relation in question is convertible, this fact achieves verbal recognition, as in the proposition "Peleus and Thetis were the parents of Achilles".[1] But not always. For example, it would scarcely be possible to put the proposition "the Earth is bigger than the Moon" into other words so as to make "the Earth and the Moon" appear as a composite subject; the "and" must always indicate that the two things are being put in some way on a level. However, this does not affect the issue.

The doctrine of relation-concepts is thus, like that of simple concepts, a part of pure logic. What is of concern to logic is not the special content of any particular relation, but only the logical form. And whatever can be asserted of this, is true analytically and known a priori. This is as true of relation-concepts as of other concepts.

Just as
$$\text{"}a \text{ falls under the concept } F\text{"}$$
is the general form of a judgement-content which deals with an object a, so we can take
$$\text{"}a \text{ stands in the relation } \phi \text{ to } b\text{"}$$
as the general form of a judgement-content which deals with an object a and an object b.

§ 71. If now every object which falls under the concept F stands in the relation ϕ to an object falling under the concept G, and if to every object which falls under G there stands in the relation ϕ an object falling under F, then the objects falling under F and under G are correlated with each other by the relation ϕ.

It may still be asked, what is the meaning of the expression
$$\text{"every object which falls under } F \text{ stands in the relation}$$
$$\phi \text{ to an object falling under } G\text{"}$$
in the case where no object at all falls under F. I understand this expression as follows:

---

[1] This type of case should not be confused with another, in which the "and" joins the subjects in appearance only, but in reality joins two propositions.

the two propositions

"*a* falls under *F*"

and

"*a* does not stand in the relation ϕ to any object
falling under *G*"

cannot, whatever be signified by *a*, both be true together;
so that either the first proposition is false, or the second is, or
both are. From this it can be seen that the proposition "every
object which falls under *F* stands in the relation ϕ to an
object falling under *G*" is, in the case where there is no object
falling under *F*, true; for in that case the first proposition

"*a* falls under *F*"

is always false, whatever *a* may be.

In the same way the proposition

"to every object which falls under *G* there stands in
the relation ϕ an object falling under *F*"

means that the two propositions

"*a* falls under *G*"

and

"no object falling under *F* stands to *a* in the
relation ϕ"

cannot, whatever *a* may be, both be true together.

§ 72. We have thus seen when the objects falling under
the concepts *F* and *G* are correlated with each other by the
relation ϕ. But now in our case, this correlation has to be
one-one. By this I understand that the two following
propositions both hold good:

1. If *d* stands in the relation ϕ to *a*, and if *d* stands in the
   relation ϕ to *e*, then generally, whatever *d*, *a* and *e* may
   be, *a* is the same as *e*.
2. If *d* stands in the relation ϕ to *a*, and if *b* stands in the
   relation ϕ to *a*, then generally, whatever *d*, *b* and *a* may be,
   d is the same as *b*.

This reduces one-one correlation to purely logical

relationships, and enables us to give the following definition:

the expression

"the concept F is equal to the concept G"

is to mean the same as the expression

"there exists a relation ϕ which correlates one to one the objects falling under the concept F with the objects falling under the concept G".

We now repeat our original definition:

the Number which belongs to the concept F is the extension of the concept "equal to the concept F"

and add further:

the expression

"n is a Number"

is to mean the same as the expression

"there exists a concept such that n is the Number which belongs to it".

Thus the concept of Number receives its definition, apparently, indeed, in terms of itself, but actually without any fallacy, since "the Number which belongs to the concept F" has already been defined.

§ 73. Our next aim must be to show that the Number which belongs to the concept F is identical with the Number which belongs to the concept G if the concept F is equal to the concept G. This sounds, of course, like a tautology. But it is not; the meaning of the word "equal" is not to be inferred from its etymology, but taken to be as I defined it above.

On our definition [of "the Number which belongs to the concept F"], what has to be shown is that the extension of the concept "equal to the concept F" is the same as the extension of the concept "equal to the concept G", if the concept F is equal to the concept G. In other words: it is to be proved that, for F equal to G, the following two propositions hold good universally:

if the concept H is equal to the concept F,

then it is also equal to the concept G;

and

if the concept H is equal to the concept G,
then it is also equal to the concept F.

The first proposition amounts to this, that there exists a relation which correlates one to one the objects falling under the concept H with those falling under the concept G, if there exists a relation ϕ which correlates one to one the objects falling under the concept F with those falling under the concept G and if there exists also a relation ψ which correlates one to one the objects falling under the concept H with those falling under the concept F. The following arrangement of letters will make this easier to grasp:

$$H \psi F \phi G.$$

Such a relation can in fact be given: it is to be found in the judgement-content

"there exists an object to which c stands in the relation ψ and which stands to b in the relation ϕ",

if we subtract from it c and b—take them, that is, as the terms of the relation. It can be shown that this relation is one-one, and that it correlates the objects falling under the concept H with those falling under the concept G.

A similar proof can be given of the second proposition also.[1] And with that, I hope, enough has been indicated of my methods to show that our proofs are not dependent at any point on borrowings from intuition, and that our definitions can be used to some purpose.

§ 74. We can now pass on to the definitions of the individual numbers.

---

[1] And likewise of the converse: If the number which belongs to the concept F is the same as that which belongs to the concept G, then the concept F is equal to the concept G.

Since nothing falls under the concept "not identical with itself", I define nought as follows:

> o is the Number which belongs to the concept "not identical with itself".

Some may find it shocking that I should speak of a concept in this connexion. They will object, very likely, that it contains a contradiction and is reminiscent of our old friends the square circle and wooden iron. Now I believe that these old friends are not so black as they are painted. To be of any use is, I admit, the last thing we should expect of them; but at the same time, they cannot do any harm, if only we do not assume that there is anything which falls under them—and to that we are not committed by merely using them. That a concept contains a contradiction is not always obvious without investigation; but to investigate it we must first possess it and, in logic, treat it just like any other. All that can be demanded of a concept from the point of view of logic and with an eye to rigour of proof is only that the limits to its application should be sharp, that it should be determined, with regard to every object whether it falls under that concept or not. But this demand is completely satisfied by concepts which, like "not identical with itself", contain a contradiction; for of every object we know that it does not fall under any such concept.[1]

On my use of the word "concept",

> "a falls under the concept F"

is the general form of a judgement-content which deals with

---

[1] The definition of an object in terms of a concept under which it falls is a very different matter. For example, the expression "the largest proper fraction" has no content, since the definite article claims to refer to a definite object. On the other hand, the concept "fraction smaller than 1 and such that no fraction smaller than one exceeds it in magnitude" is quite unexceptionable: in order, indeed, to prove that there exists no such fraction, we must make use of just this concept, despite its containing a contradiction. If, however, we wished to

an object *a* and permits of the insertion for *a* of anything whatever. And in this sense

"*a* falls under the concept 'not identical with itself' "

has the same meaning as

"*a* is not identical with itself"

or

"*a* is not identical with *a*".

I could have used for the definition of nought any other concept under which no object falls. But I have made a point of choosing one which can be proved to be such on purely logical grounds; and for this purpose "not identical with itself" is the most convenient that offers, taking for the definition of "identical" the one from LEIBNIZ given above [(§ 65)], which is in purely logical terms.

§ 75. Now it must be possible to prove, by means of what has already been laid down, that every concept under which no object falls is equal to every other concept under which no object falls, and to them alone; from which it follows that o is the Number which belongs to any such concept, and that no object falls under any concept if the number which belongs to that concept is o.

If we assume that no object falls under either the concept *F* or the concept *G*, then in order to prove them equal we have to find a relation ϕ which satisfies the following conditions: ·

every object which falls under *F* stands in the relation ϕ to an object which falls under *G*; and to every object which falls under *G* there stands in the relation ϕ an object falling under *F*.

use this concept for defining an object falling under it, it would, of course, be necessary first to show two distinct things:

1. that some object falls under this concept;
2. that only one object falls under it.

Now since the first of these propositions, not to mention the second, is false, it follows that the expression "the largest proper fraction" is senseless.

In view of what has been said above [(§ 71)] on the meaning of these expressions, it follows, on our assumption [that no object falls under either concept], that these conditions are satisfied by every relation whatsoever, and therefore among others by identity, which is moreover a one-one relation; for it meets both the requirements laid down [in § 72] above.

If, to take the other case, some object, say a, does fall under G, but still none falls under F, then the two propositions

"a falls under G"
and
"no object falling under F stands to a in the
relation ϕ"

are both true together for every relation ϕ; for the first is made true by our first assumption and the second by our second assumption. If, that is, there exists no object falling under F, then a fortiori there exists no object falling under F which stands to a in any relation whatsoever. There exists, therefore, no relation by which the objects falling under F can be correlated with those falling under G so as to satisfy our definition [of equality], and accordingly the concepts F and G are unequal.

§ 76. I now propose to define the relation in which every two adjacent members of the series of natural numbers stand to each other. The proposition:

> "there exists a concept F, and an object falling under it x, such that the Number which belongs to the concept F is n and the Number which belongs to the concept 'falling under F but not identical with x' is m"

is to mean the same as

> "n follows in the series of natural numbers directly after m".

I avoid the expression "n is *the* Number following next after m", because the use of the definite article cannot be justified until we have first proved two propositions.[1] For

---

[1] See note on p. 87e f.

the same reason I do not yet say at this point "$n = m + 1$," for to use the symbol $=$ is likewise to designate $(m + 1)$ an object.

§ 77. Now in order to arrive at the number 1, we have first of all to show that there is something which follows in the series of natural numbers directly after o.

Let us consider the concept—or, if you prefer it, the predicate—"identical with o". Under this falls the number o. But under the concept "identical with o but not identical with o", on the other hand, no object falls, so that o is the Number which belongs to this concept. We have, therefore, a concept "identical with o" and an object falling under it o, of which the following propositions hold true:

> the Number which belongs to the concept "identical with o" is identical with the Number which belongs to the concept "identical with o";
> the Number which belongs to the concept "identical with o but not identical with o" is o.

Therefore, on our definition [(§ 76)], the Number which belongs to the concept "identical with o" follows in the series of natural numbers directly after o.

Now if we give the following definition:

> 1 is the Number which belongs to the concept "identical with o",

we can then put the preceding conclusion thus:

> 1 follows in the series of natural numbers directly after o.

It is perhaps worth pointing out that our definition of the number 1 does not presuppose, for its objective legitimacy, any matter of observed fact.[1] It is easy to get confused over this, seeing that certain subjective conditions must be satisfied if we are to be able to arrive at the definition, and that sense experiences are what prompt us to frame it.[2] All this, how-

---

[1] Non-general proposition.
[2] Cf. B. Erdmann, *Die Axiome der Geometrie*, p. 164.

ever, may be perfectly correct, without the propositions so arrived at ceasing to be a priori. One such condition is, for example, that blood of the right quality must circulate in the brain in sufficient volume—at least so far as we know; but the truth of our last proposition does not depend on this; it still holds, even if the circulation stops; and even if all rational beings were to take to hibernating and fall asleep simultaneously, our proposition would not be, say, cancelled for the duration, but would remain quite unaffected. For a proposition to be true is just not the same thing as for it to be thought.

§ 78. I proceed to give here a list of several propositions to be proved by means of our definitions. The reader will easily see for himself in outline how this can be done.

1. If a follows in the series of natural numbers directly after o, then a is $= 1$.

2. If 1 is the Number which belongs to a concept, then there exists an object which falls under that concept.

3. If 1 is the Number which belongs to a concept F; then, if the object x falls under the concept F and if y falls under the concept F, x is $= y$; that is, x is the same as y.

4. If an object falls under the concept F, and if it can be inferred generally from the propositions that x falls under the concept F and that y falls under the concept F that x is $= y$, then 1 is the Number which belongs to the concept F.

5. The relation of m to n which is established by the proposition:

"n follows in the series of natural numbers directly after m"

is a one-one relation.

There is nothing in this so far to state that for every Number there exists another Number which follows directly after it, or after which it directly follows, in the series of natural numbers.

6. Every Number except o follows in the series of natural numbers directly after a Number.

§ 79. Now in order to prove that after every Number (*n*) in the series of natural numbers a Number directly follows, we must produce a concept to which this latter Number belongs. For this we shall choose the concept

"member of the series of natural numbers ending with *n*",

which requires first to be defined.

To start with, let me repeat in slightly different words the definition of following in a series given in my *Begriffs-schrift* [*Concept Writing*]*:

The proposition

'if every object to which *x* stands in the relation ϕ falls under the concept *F*, and if from the proposition that *d* falls under the concept *F* it follows universally, whatever *d* may be, that every object to which *d* stands in the relation ϕ falls under the concept *F*, then *y* falls under the concept *F*, whatever concept *F* may be"

is to mean the same as

"*y* follows in the ϕ-series after *x*"

and again the same as

"*x* comes in the ϕ-series before *y*".

§ 80. It will not be time wasted to make a few comments on this. First, since the relation ϕ has been left indefinite, the series is not necessarily to be conceived in the form of a spatial and temporal arrangement, although these cases are not excluded.

Next, there may be those who will prefer some other definition as being more natural, as for example the following: if starting from *x* we transfer our attention continually from one object to another to which it stands in the relation ϕ, and if by this procedure we can finally reach *y*, then we say that *y* follows in the ϕ-series after *x*.

* [Cp. § 91 and notes.]

Now this describes a way of discovering that y follows, it does not define what is meant by y's following. Whether, as our attention shifts, we reach y may depend on all sorts of subjective contributory factors, for example on the amount of time at our disposal or on the extent of our familiarity with the things concerned. Whether y follows in the ϕ-series after x has in general absolutely nothing to do with our attention and the circumstances in which we transfer it; on the contrary, it is a question of fact, just as much as it is a fact that a green leaf reflects light rays of certain wave-lengths whether or not these fall into my eye and give rise to a sensation, and a fact that a grain of salt is soluble in water whether or not I drop it into water and observe the result, and a further fact that it remains still soluble even when it is utterly impossible for me to make any experiment with it.

My definition lifts the matter onto a new plane; it is no longer a question of what is subjectively possible but of what is objectively definite. For in literal fact, that one proposition follows from certain others is something objective, something independent of the laws that govern the movements of our attention, and something to which it is immaterial whether we actually draw the conclusion or not. What I have provided is a criterion which decides in every case the question Does it follow after?, wherever it can be put; and however much in particular cases we may be prevented by extraneous difficulties from actually reaching a decision, that is irrelevant to the fact itself.

We have no need always to run through all the members of a series intervening between the first member and some given object, in order to ascertain that the latter does follow after the former. Given, for example, that in the ϕ-series b follows after a and c after b, then we can deduce from our definition that c follows after a, without even knowing the intervening members of the series.

Only by means of this definition of following in a series is it possible to reduce the argument from n to $(n + 1)$, which on the face of it is peculiar to mathematics, to the general laws of logic.

§ 81. If now we have for our relation ϕ the relation of m to n established by the proposition

"n follows in the series of natural numbers directly after m,"

then we shall say instead of "ϕ-series" "series of natural numbers".

I add the following further definition:

The proposition

"y follows in the ϕ-series after x or y is the same as x"

is to mean the same as

"y is a member of the ϕ-series beginning with x"

and again the same as

"x is a member of the ϕ-series ending with y".

It follows that a is a member of the series of natural numbers ending with n, if n either follows in the series of natural numbers after a or is identical with a.[1]

§ 82. It is now to be shown that—subject to a condition still to be specified—the Number which belongs to the concept

"member of the series of natural numbers ending with n"

follows in the series of natural numbers directly after n. And in thus proving that there exists a Number which follows in the series of natural numbers directly after n, we shall have proved at the same time that there is no last member of this series. Obviously this proposition cannot be established on empirical lines or by induction.

To give the proof in full here would take us too far afield. I can only indicate briefly the way it goes. It is to be proved that

1. if a follows in the series of natural numbers directly after d, and if it is true of d that:

---

[1] If n is not a Number, then n itself is the only member of the series of natural numbers ending with n,—if that is not too shocking a way of putting it.

the Number which belongs to the concept

"member of the series of natural numbers ending with *d*"

follows in the series of natural numbers directly after *d*,

then it is also true of *a* that:

the Number which belongs to the concept

"member of the series of natural numbers ending with *a*"

follows in the series of natural numbers directly after *a*.

It is then to be proved, secondly, that what is asserted of *d* and of *a* in the propositions just stated holds for the number o. And finally it is to be deduced that it also holds for *n* if *n* is a member of the series of natural numbers beginning with o. The argument here is an application of the definition I have given [(§§ 79, 81)] of the expression

"*y* follows in the series of natural numbers after *x*",

taking for our concept *F* what is asserted above [in 1.] of *d* and *a* conjointly, but with o and *n* substituted for *d* and *a*.

§ 83. In order to prove the proposition 1. of the last paragraph, we must show that *a* is the Number which belongs to the concept "member of the series of natural numbers ending with *a*, but not identical with *a*". And for this, again, it is necessary to prove that this concept has an extension identical with that of the concept "member of the series of natural numbers ending with *d*". For this we need the proposition that no object which is a member of the series of natural numbers beginning with o can follow in the series of natural numbers after itself. And this must once again be proved by means of our definition of following in a series, on the lines indicated above.[1]

---

[1] E. Schröder (op. cit., p. 63) seems to regard this proposition as a consequence of a system of notation which could conceivably be different. Here once more we must be struck by the drawback which vitiates his whole treat-

It is this that obliges us to attach a condition to the proposition that the Number which belongs to the concept

"member of the series of natural numbers ending with n"

follows in the series of natural numbers directly after n,—the condition, namely, that n must be a member of the series of natural numbers beginning with o. For this there is a convenient abbreviation, which I define as follows:

the proposition

"n is a member of the series of natural numbers beginning with o"

is to mean the same as

"n is a finite Number".

We can thus formulate the last proposition above as follows: no finite Number follows in the series of natural numbers after itself.

Infinite Numbers.

§ 84. Contrasted with the finite Numbers are the infinite Numbers. The Number which belongs to the concept "finite Number" is an infinite Number. Let us symbolize it by, say, ∞_1. If it were a finite Number, it could not follow in the series of natural numbers after itself. But it can be shown that this is what ∞_1 does.

About the infinite Number ∞_1 so defined there is nothing mysterious or wonderful. "The Number which belongs to the concept F is ∞_1" means no more and no less than this: that there exists a relation which correlates one to one the objects falling under the concept F with the finite Numbers. In terms

ment of this matter,—that we do not really know whether the number is a symbol and if so what its meaning is, or whether the number itself is the meaning of the symbol. He is not entitled to infer, from the fact that we arrange for our symbols to differ so that the same one never recurs, that the meanings of those symbols are therefore also different.

of our definitions this has a perfectly clear and unambiguous sense; and that is enough to justify the use of the symbol ∞ and to assure it of a meaning. That we cannot form any idea of an infinite Number is of absolutely no importance; the same is equally true of finite Numbers. So regarded, our Number ∞_1, has a character as definite as that of any finite Number; it can be recognized again beyond doubt as the same, and can be distinguished from every other.

§ 85. It is only recently that infinite Numbers have been introduced, in a remarkable work by G. CANTOR.[1] I heartily share his contempt for the view that in principle only finite Numbers ought to be admitted as actual. Perceptible by the senses these are not, nor are they spatial—any more than fractions are, or negative numbers, or irrational or complex numbers; and if we restrict the actual to what acts on our senses or at least produces effects which may cause sense-perceptions as near or remote consequences, then naturally no number of any of these kinds is actual. But it is also true that we have no need at all to appeal to any such sense-perceptions in proving our theorems. Any name or symbol that has been introduced in a logically unexceptionable manner can be used in our enquiries without hesitation, and here our Number ∞_1 is as sound as 2 or 3.

While in this I agree, as I believe, with CANTOR, my terminology diverges to some extent from his. For my Number he uses "power", while his concept[2] of Number has reference to arrangement in an order. Finite Numbers,

[1] Op. cit., p. 74e above.

[2] This expression may seem to conflict with my earlier insistence on the objective nature of concepts; but all that I mean is subjective here is his use of the *word*.

certainly, emerge as independent nevertheless of sequence in series, but not so transfinite Numbers. But now in ordinary use the word "Number" and the question "how many?" have no reference to arrangement in a fixed order. CANTOR's Number gives rather the answer to the question: "the how-manyeth member in the succession is the last member?" So that it seems to me that my terminology accords better with ordinary usage. If we extend the meaning of a word, we should take care that, so far as possible, no general proposition is invalidated in the process, especially one so fundamental as that which asserts of Number its independence of sequence in series. For us, because our concept of Number has from the outset covered infinite numbers as well, no extension of its meaning has been necessary at all.

§ 86. To obtain his infinite Numbers CANTOR introduces the relation-concept of following in a succession, which differs from my "following in a series". On his account we should get a succession if, for example, we arranged the finite positive whole numbers in an order such that the odd numbers followed one another just as they do, among themselves, in the series of natural numbers, and similarly the even numbers, but with the further stipulation that every even number was to follow after every odd number. In this succession o, for instance, would follow after 13. But no number would come directly before o. Now this is a situation which cannot arise on my definition of following in the series. It can be strictly proved, without appeal to any axiom borrowed from intuition, that if y follows in the ϕ-series after x then there exists an object which comes in that series directly before y. Now it looks to me as though precise definitions of following in the succession and of Number in CANTOR's sense are still wanting. Thus CANTOR appeals to the rather mysterious "inner intuition", where he ought to have made an effort to find, and indeed could actually have found, a proof from definitions. For I think I can anticipate how his two concepts could have been made precise. At any rate, nothing in what I have said is intended to question in any way their legitimacy or their fertility. On the contrary, I find special reason to welcome in CANTOR's investigations an extension of the frontiers of science, because they have led to the construction of a purely arithmetical route to higher transfinite Numbers (powers).

V: Conclusion.

§ 87. I hope I may claim in the present work to have made it probable that the laws of arithmetic are analytic judgements and consequently a priori. Arithmetic thus becomes simply a development of logic, and every proposition of arithmetic a law of logic, albeit a derivative one. To apply arithmetic in the physical sciences is to bring logic to bear on observed facts;[1] calculation becomes deduction. The laws of number will not, as BAUMANN[2] thinks, need to stand up to practical tests if they are to be applicable to the external world; for in the external world, in the whole of space and all that therein is, there are no concepts, no properties of concepts, no numbers. The laws of number, therefore, are not really applicable to external things; they are not laws of nature. They are, however, applicable to judgements holding good of things in the external world: they are laws of the laws of nature. They assert not connexions between phenomena, but connexions between judgements; and among judgements are included the laws of nature.

§ 88. KANT[3] obviously—as a result, no doubt, of defining them too narrowly—underestimated the value of analytic judgements, though it seems that he did have some inkling

---

[1] Observation itself already includes within it a logical activity.
[2] Op. cit., Vol. II, p. 670.
[3] Op. cit., Vol. III, pp. 39 ff. [Original edns., A6 ff./B10 ff.].

of the wider sense in which I have used the term.[1] On the basis of his definition, the division of judgements into analytic and synthetic is not exhaustive. What he is thinking of is the universal affirmative judgement; there, we can speak of a subject concept and ask—as his definition requires—whether the predicate concept is contained in it or not. But how can we do this, if the subject is an individual object? Or if the judgement is an existential one? In these cases there can simply be no question of a subject concept in KANT's sense. He seems to think of concepts as defined by giving a simple list of characteristics in no special order; but of all ways of forming concepts, that is one of the least fruitful. If we look through the definitions given in the course of this book, we shall scarcely find one that is of this description. The same is true of the really fruitful definitions in mathematics, such as that of the continuity of a function. What we find in these is not a simple list of characteristics; every element in the definition is intimately, I might almost say organically, connected with the others. A geometrical illustration will make the distinction clear to intuition. If we represent the concepts (or their extensions) by figures or areas in a plane, then the concept defined by a simple list of characteristics corresponds to the area common to all the areas representing the defining characteristics; it is enclosed by segments of their boundary lines. With a definition like this, therefore, what we do—in terms of our illustration—is to use the lines already given in a new way for the purpose of demarcating an area.[2] Nothing essentially new, however, emerges in the process. But the more fruitful type of definition is a matter of drawing boundary lines that were not previously given at all. What we shall be able to infer from it, cannot be inspected in advance; here,

[1] On p. 43 [B14] he says that a synthetic proposition can only be seen to be true by the law of contradiction, if another synthetic proposition is presupposed.

[2] Similarly, if the characteristics are joined by "or".

we are not simply taking out of the box again what we have just put into it. The conclusions we draw from it extend our knowledge, and ought therefore, on KANT's view, to be regarded as synthetic; and yet they can be proved by purely logical means, and are thus analytic. The truth is that they are contained in the definitions, but as plants are contained in their seeds, not as beams are contained in a house. Often we need several definitions for the proof of some proposition, which consequently is not contained in any one of them alone, yet does follow purely logically from all of them together.

§ 89. I must also protest against the generality of KANT's[1] dictum: without sensibility no object would be given to us. Nought and one are objects which cannot be given to us in sensation. And even those who hold that the smaller numbers are intuitable, must at least concede that they cannot be given in intuition any of the numbers greater than 1000 1000 1000, about which nevertheless we have plenty of information. Perhaps KANT used the word "object" in a rather different sense; but in that case he omits altogether to allow for nought or one, or for our ∞_1,—for these are not concepts either, and even of a concept KANT requires that we should attach its object to it in intuition.

I have no wish to incur the reproach of picking petty quarrels with a genius to whom we must all look up with grateful awe; I feel bound, therefore, to call attention also to the extent of my agreement with him, which far exceeds any disagreement. To touch only upon what is immediately relevant, I consider KANT did great service in drawing the distinction between synthetic and analytic judgements. In calling the truths of geometry synthetic and a priori, he

[1] Op. cit., Vol. III, p. 82 [Original edns., A51/B75.]

revealed their true nature. And this is still worth repeating, since even to-day it is often not recognized. If KANT was wrong about arithmetic, that does not seriously detract, in my opinion, from the value of his work. His point was, that there are such things as synthetic judgements a priori; whether they are to be found in geometry only, or in arithmetic as well, is of less importance.

§ 90. I do not claim to have made the analytic character of arithmetical propositions more than probable, because it can still always be doubted whether they are deducible solely from purely logical laws, or whether some other type of premiss is not involved at some point in their proof without our noticing it. This misgiving will not be completely allayed even by the indications I have given of the proof of some of the propositions; it can only be removed by producing a chain of deductions with no link missing, such that no step in it is taken which does not conform to some one of a small number of principles of inference recognized as purely logical. To this day, scarcely one single proof has ever been conducted on these lines; the mathematician rests content if every transition to a fresh judgement is self-evidently correct, without enquiring into the nature of this self-evidence, whether it is logical or intuitive. A single such step is often really a whole compendium, equivalent to several simple inferences, and into it there can still creep along with these some element from intuition. In proofs as we know them, progress is by jumps, which is why the variety of types of inference in mathematics appears to be so excessively rich; for the bigger the jump, the more diverse are the combinations it can represent of simple inferences with axioms derived from intuition. Often, nevertheless, the correctness of such a transition is immediately self-evident to us, without our ever becoming conscious of the subordinate steps condensed within it; whereupon, since it does not obviously conform to any of the recognized types of logical inference, we are prepared to accept its self-evidence forthwith as intuitive, and the conclusion itself as a synthetic truth—and this even when obviously it holds good of much more than merely what can be intuited.

On these lines what is synthetic and based on intuition cannot be sharply separated from what is analytic. Nor shall we succeed in compiling with certainty a complete set of axioms of intuition, such that from them alone we can derive, by means of the laws of logic, every proof in mathematics.

§ 91. The demand is not to be denied: every jump must be barred from our deductions. That it is so hard to satisfy must be set down to the tediousness of proceeding step by step. Every proof which is even a little complicated threatens to become inordinately long. And moreover, the excessive variety of logical forms that have been developed in our language makes it difficult to isolate a set of modes of inference which is both sufficient to cope with all cases and easy to take in at a glance.

To minimize these drawbacks, I invented my concept writing. It is designed to produce expressions which are shorter and easier to take in, and to be operated like a calculus by means of a small number of standard moves, so that no step is permitted which does not conform to the rules which are laid down once and for all.[1] It is impossible, therefore, for any premiss to creep into a proof without being noticed. In this way I have, without borrowing any axiom from intuition, given a proof of a proposition[2] which might at first sight be taken for synthetic, which I shall here formulate as follows:

If the relation of every member of a series to its successor is (one- or) many-one, and if m and y follow in that series after x, then either y comes in that series before m, or it coincides with m, or it follows after m.

---

[1] It is designed, however, to be capable of expressing not only the logical form, like Boole's notation, but also the content of a proposition.

[2] *Begriffsschrift*, Halle a/S. 1879, p. 86, Formula 133.

From this proof it can be seen that propositions which extend our knowledge can have analytic judgements for their content.[1]

Other numbers.

§ 92. Up to now we have restricted our treatment to the [natural] Numbers. Let us now take a look at the other kinds of numbers, and try to make some use in this wider field of what we have learned in the narrower.

HANKEL,[2] in an attempt to make clear the sense of asking whether some particular type of number is possible, writes as follows:

"Number to-day is no longer a thing, a substance, existing in its own right apart from the thinking subject and the objects which give rise to it, a self-subsistent element in the sort of way it was for the Pythagoreans. The question whether some number exists can therefore only be understood as referring to the thinking subject or to the objects thought about, relations between which the numbers represent. As impossible in the strict sense the mathematician counts only what is logically impossible, that is, self-contradictory. That numbers which are impossible in this sense cannot be admitted, needs no proof. But if the numbers concerned are logically possible, if their concept is clearly and fully defined and there-

---

[1] This proof will certainly still be found far too lengthy, a disadvantage which may, perhaps, be thought to be more than outweighed by the practically absolute certainty that it contains no mistake and no gap. My aim at that time was to reduce everything to the smallest possible number of the simplest possible logical laws. Consequently, I made use of only one principle of deduction. However, even at that time I noted in my Preface, p. vii, that for the further application of my writing it would be imperative to admit more such principles. This can be done without loosening any link in the chain of deduction, and it is possible to achieve in this way a remarkable degree of compression.

[2] Op. cit., pp. 6–7.

fore free from contradiction, then the question whether they exist can amount only to this: Does there exist in reality or in the actual world given to us in intuition a substratum for these numbers, do there exist objects in which they—relations, that is, for the mind, of the type defined—can become phenomenal?"

§ 93. HANKEL's first sentence leaves it doubtful whether he holds that numbers exist in the thinking subject, or in the objects which give rise to them, or in both. In the spatial sense they are, in any case, neither inside nor outside either the subject or any object. But, of course, they are outside the subject in the sense that they are not subjective. Whereas each individual can feel only his own pain or desire or hunger, and can experience only his own sensations of sound and colour, numbers can be objects in common to many individuals, and they are in fact precisely the same for all, not merely more or less similar mental states in different minds. In making the question of the existence of numbers refer to the thinking subject, HANKEL seems to make it a psychological question, which it is not in any way. Mathematics is not concerned with the nature of our mind, and the answer to any question whatsoever in psychology must be for mathematics a matter of complete indifference.

§ 94. Further, exception must be taken to the statement that the mathematician counts as impossible only what is self-contradictory. A concept is still admissible even though its defining characteristics do contain a contradiction: all that we are forbidden to do, is to presuppose that something falls under it. But even if a concept contains no contradiction, we still cannot infer that for that reason something falls under it. If such concepts were not admissible, how could we ever prove that a concept does not contain any contradiction? It is by no means always obvious; it does not follow that because we see no contradiction there is none there, nor does a clear and full definition afford any guarantee against it. HANKEL[1] proves that any closed field of complex numbers of higher order than the ordinary, if made subject to all the laws of addition and multiplication, contains a contradiction. Now that is something that needs to be proved; it is not seen immediately. Before his proof was given, anyone could always, by using a number system of that type, have arrived at remarkable results, nor would they have been any worse founded than the theory of determinants, if, with HANKEL,[2] we base that on alternate numbers; for who can assure us that there is not some hidden contradiction in the concept of these numbers also? And moreover, even if we could exclude this possibility generally for as many alternate units as we please, it would still not follow that such units exist. Yet that is precisely what we need. We will take an example from EUCLID's *Elements*, Book I, Theorem 18:

> In any triangle the greater side subtends the greater angle.

To prove this, EUCLID cuts off from the greater side AC a segment AD equal to the lesser side AB, making use for this purpose of a previously given construction. The proof would collapse, if there were no such point as D, and it is not enough that we discover no contradiction in the concept "point on AC whose distance from A is equal to B's". EUCLID proceeds to join BD. That there exists such a line is still another proposition on which the proof depends.

§ 95. Strictly, of course, we can only establish that a concept is free from contradiction by first producing something that falls under it. The converse inference is a fallacy, and one into which HANKEL falls. Referring to [the operation of finding x from] the equation $x + b = c$ [(subtraction)] he says:[3]

[1] Op. cit., pp. 106-7.
[2] Op. cit., § 35, pp. 121-4.
[3] Op. cit., p. 5. Similarly E. Kossak, op. cit., p. 17 *ad fin.*

"It is obvious that, for $b > c$, there is no number x in the series 1, 2, 3, . . . which solves our problem; the subtraction is then *impossible*. There is nothing, however, to prevent us from regarding the difference $(c - b)$ in this case as a *symbol* which solves the problem and which is to be operated with exactly as if it were a figure number in the series 1, 2, 3 . . ."

Nevertheless, there is something to prevent us from regarding $(2 - 3)$ without more ado as a symbol which solves the problem; for an empty symbol is precisely no solution; without some content it is merely ink or print on paper, as which it possesses physical properties but not that of making 2 when increased by 3. Really, it would not be a symbol at all, and to use it as one would be a mistake in logic. Even for $c > b$, it is not the symbol ("$c - b$") that solves the problem, but its content.

§ 96. We might just as well say this: among numbers hitherto known there is none which satisfies the simultaneous equations

$$x + 1 = 2$$
$$x + 2 = 1,$$

but there is nothing to prevent us from introducing a symbol which solves the problem. Ah, but there is, it will be replied: to satisfy both the equations simultaneously involves a contradiction. Certainly, if we are requiring a real number or an ordinary complex number to satisfy them; but then all we have to do is to widen our number system, to create numbers which do meet these new requirements. Then we can wait and see whether anyone succeeds in producing a contradiction in them. Who can tell what may not be possible with our new numbers? Naturally $(c - b)$ cannot then remain one-valued; but then $\sqrt{a}$ likewise, if we wish to introduce negative numbers, has to cease to be one-valued; and with complex numbers $\log a$ too becomes many-valued.

And why not create still further numbers which permit the summation of diverging series? But that will do,—even the mathematician cannot create things at will, any more than the geographer can; he too can only discover what is there and give it a name.

This is the error that infects the formalist theory of fractions and of negative and complex numbers.[1] It is made a postulate that the familiar rules of calculation shall still hold, where possible, for the newly-introduced numbers, and from this their general properties and relations are deduced. If no contradiction is anywhere encountered, the introduction of the new numbers is held to be justified, as though it were impossible for a contradiction still to be lurking somewhere nevertheless, and as though freedom from contradiction amounted straight away to existence.

§ 97. That this mistake is so easily made is due, of course, to the failure to distinguish clearly between concepts and objects. Nothing prevents us from using the concept "square root of -1"; but we are not entitled to put the definite article in front of it without more ado and take the expression "the square root of -1" as having a sense. Given that $i^2 = -1$, we can give a proof of the formula expressing the sine of any multiple of the angle a in terms of $\sin a$ and $\cos a$; but we ought not to forget that this proposition continues to imply the condition that $i^2 = -1$, which we are not entitled to drop without remark. If there existed nothing at all of which the square was -1, then for all our proof was worth the formula might not be correct,[2] since the condition $i^2 = -1$, on which its validity patently depends, would never be fulfilled. It would be as though in a geometrical proof we had made use of an auxiliary line which is quite impossible to construct.

§ 98. HANKEL[3] introduces two sorts of operation, which he calls lytic and thetic, and which he defines by means of

[1] CANTOR's infinite Numbers are in like case.
[2] It might always be possible to prove it strictly in some other way.
[3] Op. cit., p. 18.

certain properties that they are to possess. There is nothing against this, so long as it is only not presupposed that operations of these sorts and objects such as their results would be exist.[1] Later[2] he symbolizes an operation which is thetic, one-valued* and associative by $(a + b)$, and the corresponding lytic, and likewise one-valued*, operation by $(a - b)$. *An operation which etc.?* But which one? Any we care to choose? Then that is not a definition of $(a + b)$; and besides, what if none such exists? If the word "addition" had as yet no meaning, it would be quite in order logically to say: we propose to call an operation of this sort an addition; but what we cannot say is: we propose to call an operation of this sort *the* operation of addition, and to symbolize it by $(a + b)$. For it has not yet been established that there is one and only one such operation. We cannot define by putting on one side of our identity the indefinite article and on the other the definite. HANKEL, however, goes on to speak next without more ado of "the modulus of the operation", without having proved that there is one and only one modulus.

§ 99. In a word, this purely formalist theory is not sufficient. What is valuable in it is simply this. We can prove that if any operation possesses certain properties, such as that of being associative or commutative, then certain propositions hold good of it. So that if we go on to show that addition and multiplication, which are already known to us, possess these properties, we can then proceed immediately to assert our propositions of addition and multiplication, without repeating the proof at length for each case individually. Thus it is only after we have applied our formal theory to operations given from elsewhere, that we arrive at the familiar propositions of arithmetic. But we have not the slightest right to suppose that we can use it as a method for introducing addition and multiplication. It does not give

[1] This Hankel really does already by using the identity $\Theta(c, b) = a$.
[2] Op. cit., p. 29.

* [Literally 'perfectly one-valued', a term defined by Hankel.]

their actual definitions, but only lays down the lines for them. We may say: the name "addition" is to be given only to an operation which is thetic, one-valued and associative, but there is nothing at all in this as yet to say which operation it is that is to be so called. So far as this goes, there is nothing to stop us calling multiplication addition and symbolizing it by $(a + b)$, nor could anyone say definitely whether $2 + 3$ was 5 or 6.

§ 100. If we abandon this purely formal method of treatment, we may fasten instead on the circumstance that, simultaneously with the introduction of new numbers, the meanings of the words "sum" and "product" are extended. We take some object, let us say the Moon, and proceed by definition: Let the Moon multiplied by itself be —1. This gives us a square root of —1 in the shape of the Moon. There seems to be nothing wrong with this definition, since the meaning hitherto assigned to multiplication says nothing as to the sense of a product such as the Moon into the Moon, so that as we now come to extend its meaning we can make it, for the Moon, whatever we choose. But we need also the product of a real number into the square root of —1. So let us choose instead as our square root of —1 the time-interval of one second, and let this be symbolized by i. Thus $3i$ will mean the time-interval of 3 seconds, and so on.[1] What object shall we then symbolize by, say, $2 + 3i$? What meaning should be assigned to the plus symbol in this case? Now this must be

[1] We should be equally entitled to choose as further square roots of —1 a certain quantum of electricity, a certain surface area, and so on; but then we should naturally have to use different symbols to signify these different roots. That we are able, apparently, to create in this way as many square roots of —1 as we please, is not so astonishing when we reflect that the meaning of the square root of —1 is not something which was already unalterably fixed before we made these choices, but is decided for the first time by and along with them.

laid down generally for all such cases, which clearly is not going to be easy. However, let us just assume that we have successfully secured a sense for all symbols of the form $a + bi$, and a sense such that the familiar laws of addition hold good of it. What we should then have to do would be to lay it down further that in general

$$(a + bi)(c + di) = ac - bd + i(ad + bc),$$

thus defining the extended meaning of multiplication.

§ 101. We should now be able to prove the formula for cos $(n\,a)$, if we knew that from the identity of complex numbers the identity of their real parts can be inferred. That would have to result from the sense of $a + bi$, which we are here taking to have been made available. Our proof of the formula would thus hold only for complex numbers and their sums and products in the sense fixed by us. Now since for real integral n and real a i disappears completely from the identity in the formula, we are tempted to conclude therefore that it is quite immaterial whether i means a second or a millimetre or anything else, provided only that our laws of addition and multiplication hold good; everything depends on that, and the rest we need not bother about. Well, perhaps it is indeed possible to assign a whole variety of different meanings to $a + bi$, and to sum and product, all of them such that those laws continue to hold good; but it is not immaterial whether we can or cannot find *some* such a sense for those expressions.

§ 102. It is common to proceed as if a mere postulation were equivalent to its own fulfilment. We postulate that it shall be possible in all cases to carry out the operation of subtraction,[1] or of division, or of root extraction, and suppose that with that we have done enough. But why do we not postulate that through any three points it shall be possible to draw a straight line? Why do we not postulate that all the laws

[1] Cp. Kossak, op. cit., p. 17.

of addition and multiplication shall continue to hold for a three-dimensional complex number system just as they do for real numbers? Because this postulate contains a contradiction. Very well then, what we have to do first is to prove that these other postulates of ours do not contain any contradiction. Until we have done that, all rigour, strive for it as we will, is so much moonshine.

In a geometrical theorem where a constructed line is used for the proof, the auxiliary line does not occur in the theorem. Perhaps more than one such line is possible, as for instance where we can select a point at will. But however much we can dispense with each and any of them individually, still the cogency of our proof depends on its being possible to draw some line of the required character. Merely to postulate it is not enough. So in our case likewise, it is not immaterial to the cogency of our proof whether "$a + bi$" has a sense or is nothing more than printer's ink. It will not get us anywhere simply to require that it have a sense, or to say that it is to have the sense of the sum of a and bi, when we have not previously defined what "sum" means in this case and when we have given no justification for the use of the definite article.

§ 103. Against the particular sense we have proposed to assign to "i" many objections can of course be brought. By it, we are importing into arithmetic something quite foreign to it, namely time. The second stands in absolutely no intrinsic relation to the real numbers. Propositions proved by the aid of complex numbers would become a posteriori judgements, or rather, at any rate, synthetic, unless we could find some other sort of proof for them or some other sense for i. We must at least first make the attempt to show that all propositions of arithmetic are analytic.

Kossak's[1] account of complex number—"the composite

[1] Op. cit., p. 17.

idea of heterogeneous groups of identical elements"[1]—
appears to avoid importing anything foreign, but this appearance is only due to the vagueness of his terminology. We are given no answer at all to the question, what does $1 + i$ really mean? Is it the idea of an apple and a pear, or the idea of toothache and gout? Not both at once, at any rate, because then $1 + i$ would not be always identical with $1 + i$. The temptation is to say: it depends on the special meaning we assign to it. Very well then, KOSSAK's statement once again does not yet give us any definition at all of complex number, it only lays down the general lines to proceed along. But we need more; we must know definitely what "i" means, and if we do proceed along his lines and try saying it means the idea of a pear, we shall once again be introducing something foreign into arithmetic.

What is commonly called the geometrical representation of complex numbers has at least this advantage over the proposals so far considered, that in it 1 and i do not appear as wholly unconnected and different in kind: the segment taken to represent i stands in a regular relation to the segment which represents 1. Though I may add that, strictly, it is not correct that 1 here means a certain segment and i a segment perpendicular to it of the same length; on the contrary, "1" means in all contexts the same. A complex number, on this interpretation, shows how the segment taken as its representation is reached, starting from a given segment (the unit segment), by means of operations of multiplication, division, and rotation.[2] However, even this account seems to make every theorem whose proof has to be based on the existence of a complex number dependent on geometrical intuition and so synthetic.

[1] Cf. for the term "idea" § 27, for "group" what is said about "agglomeration" in § 23 and § 25, and for the identity of the elements §§ 34–39.
[2] For simplicity I neglect incommensurables here.

§ 104. How are complex numbers to be given to us then, and fractions and irrational numbers? If we turn for assistance to intuition, we import something foreign into arithmetic; but if we only define the concept of such a number by giving its characteristics, if we simply require the number to have certain properties, then there is still no guarantee that anything falls under the concept and answers to our requirements, and yet it is precisely on this that proofs must be based.

Well, how do things stand with the [natural] Numbers? Have we really no right to speak of $1000^{1000^{1000}}$ until such time as that many objects have been given to us in intuition? Is it, till then, an empty symbol? Not at all. It has a perfectly definite sense, even although, psychologically speaking and having regard to the shortness of human life, it is impossible for us ever to become conscious of that many objects;[1] in spite of that, $1000^{1000^{1000}}$ is still an object, whose properties we can come to know, even though it is not intuitable. To convince ourselves of this, we have only to show, introducing the symbol a^n for the nth power of a, that for positive integral a and n this expression always refers to one and only one positive whole number. To give the proof of this in detail would take us too far afield for present purposes. A general idea of the way it goes can be gathered from the method used to define nought in § 74, one in § 77, and the infinite Number ∞_1 in § 84, and from the outline of the proof that after every finite Number in the series of natural numbers a Number directly follows (§§ 82–3).

In the same way with the definitions of fractions, complex numbers and the rest, everything will in the end come down to the search for a judgement-content which can be transformed into an identity whose sides precisely are the new

[1] A simple calculation shows that millions of years would not be time enough for that.

numbers. In other words, what we must do is fix the sense of a recognition-judgement for the case of these numbers. In doing so, we must not forget the doubts raised by such transformations, which we discussed in §§ 63–68. If we follow the same procedure as we did there, then the new numbers are given to us as extensions of concepts.

§ 105. On this view of numbers[1] the charm of work on arithmetic and analysis is, it seems to me, easily accounted for. We might say, indeed, almost in the well-known words: the reason's proper study is itself. In arithmetic we are not concerned with objects which we come to know as something alien from without through the medium of the senses, but with objects given directly to our reason and, as its nearest kin, utterly transparent to it.[2]

And yet, or rather for that very reason, these objects are not subjective fantasies. There is nothing more objective than the laws of arithmetic.

§ 106. Let us cast a final brief glance back over the course of our enquiry. After establishing that number is neither a collection of things nor a property of such, yet at the same time is not a subjective product of mental processes either, we concluded that a statement of number asserts something objective of a concept. We attempted next to define the individual numbers 0, 1, etc., and the step from one number to the next in the number series. Our first attempt broke down, because we had defined only the predicate which we said was

---

[1] It too might be called formalist. However, it is completely different from the view criticized above under that name.

[2] By this I do not mean in the least to deny that without sense impressions we should be as stupid as stones, and should know nothing either of numbers or of anything else; but this psychological proposition is not of the slightest concern to us here. Because of the ever-present danger of confusing two fundamentally different questions, I make this point once more.

asserted of the concept, but had not given separate definitions of o or 1, which are only elements in such predicates. This resulted in our being unable to prove the identity of numbers. It became clear that the number studied by arithmetic must be conceived not as a dependent attribute, but substantivally.[1] Number thus emerged as an object that can be recognized again, although not as a physical or even a merely spatial object, nor yet as one of which we can form a picture by means of our imagination. We next laid down the fundamental principle that we must never try to define the meaning of a word in isolation, but only as it is used in the context of a proposition: only by adhering to this can we, as I believe, avoid a physical view of number without slipping into a psychological view of it. Now for every object there is one type of proposition which must have a sense, namely the recognition-statement, which in the case of numbers is called an identity. Statements of number too are, we saw, to be considered as identities. The problem, therefore, was now this: to fix the sense of a numerical identity, that is, to express that sense without making use of number words or the word "number". The content of a recognition-judgement concerning numbers we found to be this, that it is possible to correlate one to one the objects falling under a concept F with those falling under a concept G. Accordingly, our definition had to lay it down that a statement of this possibility means the same as a numerical identity. We recalled similar cases: the definition of direction derived from parallelism of lines, that of shape derived from similarity of figures, and so on.

§ 107. The question then arose: when are we entitled to regard a content as that of a recognition-judgement? For this a certain condition has to be satisfied, namely that it

---

[1] The distinction corresponds to that between "blue" and "the colour of the sky".

must be possible in every judgement to substitute without loss of truth the right-hand side of our putative identity for its left-hand side. Now at the outset, and until we bring in further definitions, we do not know of any other assertion concerning either side of such an identity except the one, that they are identical. We had only to show, therefore, that the substitution is possible in an identity.

One doubt, however, still remained, which was this. A recognition-statement must always have a sense. But now if we treat the possibility of correlating one to one the objects falling under the concept *F* with the objects falling under the concept *G* as an identity, by putting for it: "the Number which belongs to the concept *F* is identical with the Number which belongs to the concept *G*", thus introducing the expression "the Number which belongs to the concept *F*", this gives us a sense for the identity only if both sides of it are of the form just mentioned. A definition like this is not enough to enable us to decide whether an identity is true or false if only one side of it is of this form. We were thus led to give the definition:

The Number which belongs to the concept *F* is the extension of the concept "concept equal to the concept F", where a concept *F* is called equal to a concept *G* if there exists the possibility of one-one correlation referred to above.

In this definition the sense of the expression "extension of a concept" is assumed to be known. This way of getting over the difficulty cannot be expected to meet with universal approval, and many will prefer other methods of removing the doubt in question. I attach no decisive importance even to bringing in the extensions of concepts at all.

§ 108. It now still remained to define one-one correlation; this we reduced to purely logical relationships. Next, we first gave an outline of the proof of the proposition: the number which belongs to the concept F is identical with the number which belongs to the concept G, if the concept F is equal to the concept G; and then gave definitions of nought, of the expression "n follows in the series of natural numbers directly after m", and of the number 1, showing that 1 follows in the series of natural numbers directly after o. After adducing a number of propositions which can easily be proved at this stage, we proceeded to go rather more closely into the following proposition, from which we learn that the number series is infinite:

After every number there follows in the series of natural numbers a number.

This led us to the concept "member of the series of natural numbers ending with n", with the aim of showing that the Number belonging to this concept follows in the series of natural numbers directly after n. We began by defining this in terms of the following of an object y after an object x in a series in general ϕ. The sense of this expression too was reduced to purely logical relationships. And by this means we succeeded in showing that the inference from n to $(n + 1)$, which is ordinarily held to be peculiar to mathematics, is really based on the universal principles of inference in logic.

Now to prove that the number series is infinite, we needed to make use of the proposition that no finite number follows in the series of natural numbers after itself. And so we arrived at the concepts of finite and infinite number. We showed that fundamentally the latter is no less logically justified than the former. For the sake of comparison, CANTOR's infinite Numbers and his "following in the succession" were referred to, and at the same time the divergence in his terminology was pointed out.

§ 109. From all the preceding it thus emerged as a very probable conclusion that the truths of arithmetic are analytic and a priori; and we achieved an improvement on the view of

KANT. We saw further what is still needed to raise this probability to a certainty, and indicated the path which must lead to that goal.

Finally, we made use of our results in a critique of a formalist theory of negative, fractional, irrational and complex numbers, which made the inadequacy of the theory evident. We came to see that its error lies in taking it as proved that a concept is free from contradiction if no contradiction has revealed itself, and in taking freedom from contradiction in a concept as sufficient guarantee in itself that something falls under it. This theory imagines that all we need do is make postulates; that these are satisfied then goes without saying. It conducts itself like a god, who can create by his mere word whatever he wants. It had also to be censured for passing off as a definition what is only a guide towards a definition, and one which, if we followed it, would lead to the introduction into arithmetic of foreign elements; these do not, it is true, obtrude into the words of the "definition", but only because it remains a mere guide.

The formalists are thus in danger of relapsing into an a posteriori or at any rate a synthetic theory, however high on the summits of abstraction they may seem to themselves to be floating.

Now we, from our previous treatment of the positive whole numbers, have seen that it is possible to avoid all importation of external things and geometrical intuitions into arithmetic, without, for all that, falling into the error of the formalists. Here, just as there, it is a matter of fixing the content of a recognition-judgement. Once suppose this everywhere accomplished, and numbers of every kind, whether negative, fractional, irrational or complex, are revealed as no more mysterious than the positive whole numbers, which in turn are no more real or more actual or more palpable than they.

Printed and bound by CPI Group (UK) Ltd, Croydon, CR0 4YY

27/10/2024

14580385-0001